A phytogeographical study of the vascular plants of northern Greenland – north of 74° northern latitude

Christian Bay

MEDDELELSER OM GRØNLAND, BIOSCIENCE 36 · 1992

Contents

A phytogeographical study of the vascular plants of northern Greenland – north of 74° northern latitude

CHRISTIAN BAY

Bay, C. 1992. A phytogeographical study of the vascular plants of northern Greenland – north of 74° northern latitude. – Meddr Greenland, Biosci. 36, 102 pp. Copenhagen 1992-12-22.

The phytogeography, flora, and vegetation of northern Greenland, north of 74°N have been investigated. The study area comprises the northern part of high arctic Greenland. The distribution of the 218 taxa of vascular plants has been mapped and classified into 14 distribution types with 16 subtypes. Based on the larger plant material obtained during the intensive botanical exploration of the area in the last decade the delimitation of the phytogeographical units as defined by Seidenfaden and Sørensen (1937) and Böcher *el al.* (1959) has been tested.

The delimitation of the floristic province North Greenland, north of 79°30′N, has been confirmed, now divided, however, into a coastal and an inland district based on the distribution of selected species which in North Greenland are restricted to the interior. The coastal district is considered the polar desert zone in high arctic Greenland.

The delimitation of the districts in the floristic province Northwest Greenland has been accepted, whereas the delimitation of the originally proposed districts in the floristic province Northeast Greenland has been altered. Here the border between the continental and the oceanic floristic province has been moved in an eastward direction. Further, the northern district in the continental province is divided into two districts based on a marked floristic limit by Bessel Fjord (76°N), where several low arctic species have their northern limit, thus giving a total of four districts in continental Northeast Greenland. No evidence for maintaining the subdivision of the oceanic floristic province in Northeast Greenland at Wollaston Forland has been found. This limit is expected to be found south of the study area.

Four taxa new to the flora of Greenland have been recognized or found: *Puccinellia bruggemanni* Th. Sør., *Phippsia algida* (Sol.) R. Br. ssp. *algidiformis* (H. Sm.) L. & L., *Geum rossii* (R. Br.) Ser., and *Pedicularis sudetica* Willd. ssp. *albolabiata* Hult. giving a total of 121 taxa known from North Greenland north of 79°30′N. The total number of taxa in Northwest and Northeast Greenland north of 74°N is 161 and 194, respectively. Five different types of high arctic distribution are recognized in Greenland.

A classification of the vegetation based on detailed description of the vegetation and few analyses in Northwest, North, and Northeast Greenland is given. Two vegetation maps (scale c. 1:2, 380000) based on NOAA-satellite images showing the degree of plant cover in eastern North Greenland and Northeast Greenland, north of 75°N lat., are presented.

Christian Bay, Botanical Museum, University of Copenhagen, Gothersgade 130, DK-1123 Copenhagen K, Denmark.

1. Introduction

Until quite recently Greenland north of 79°30′N was one of the most inaccessible areas in the Arctic. The exploration of the natural history began by the end of the last century when only few expeditions succeeded in travelling along the north coast of Greenland. The only botanical information available from the area until the 1940's was from "The Second Thule Expedition" lead by Knud Rasmussen and from a few other non-botanical expeditions (Bessels 1879, Hart 1880). Thanks to the Peary Land Expeditions in 1947–50 and 1963 and especially the intensive botanical activities in the past decade it is now possible to make a comprehensive study of the flora, phytogeography, and vegetation of the floristic province North Greenland and adjacent areas in Northwest and Northeast Greenland.

The need for extending the knowledge of the basic biological elements is increasing in these years where man focuses on exploitating the natural resources in the Arctic. Floristical and phytogeographical studies are a prerequisite for planning and conducting detailed ecological studies in connection with either basic research projects e.g. vegetation mapping and foraging dynamics of herbivores, or environmental studies with the purpose of setting up regulations for minimizing the impact on vegetation and wildlife caused by human activity such as mining operations.

The present study of northern Greenland is part of a phytogeographical investigation of Greenland initiated in 1962 by the establishment of Greenland Botanical Survey of the University of Copenhagen. South Greenland has been studied by Feilberg (1984) and West Greenland between 62°20′N and 74°N is under preparation (Fredskild in prep.).

Northern Greenland is one of the largest continuous land areas of the High Arctic and the study area comprises the main part of the National Park in North and Northeast Greenland including the area from Hall Land in northwest to Jameson Land (70–71°N) in southeast. The northern part of the study area is influenced by extreme physical conditions which are reflected in the flora and vegetation. Below, the name North Greenland means the floristic province North Greenland situated north of 79°30′N (Fig. 1). The adjacent areas in the west and east will be referred to as Northwest and Northeast Greenland, respectively. These belong to the northern parts of the floristic provinces Northwest and Northeast Greenland.

2. Study area

The study area northern Greenland – north of 74°N – stretches from the middle of Melville Bugt in west around the north coast to Gael Hamkes Bugt in east extending as far north as 83°39′N. The longitudinal borders are 12°30′W in east and 73°W in west and the area includes the northernmost ice-free land in the world. It includes the floristic province North Greenland as defined by Böcher *et al.* (1959) and the northernmost parts of the adjacent provinces in West and East Greenland (Fig. 1). The southern limit is chosen at 74°N lat. in order to determine the northern limit in East Greenland of a large number of low arctic species, known to reach as far as just north of 74°N. Many of

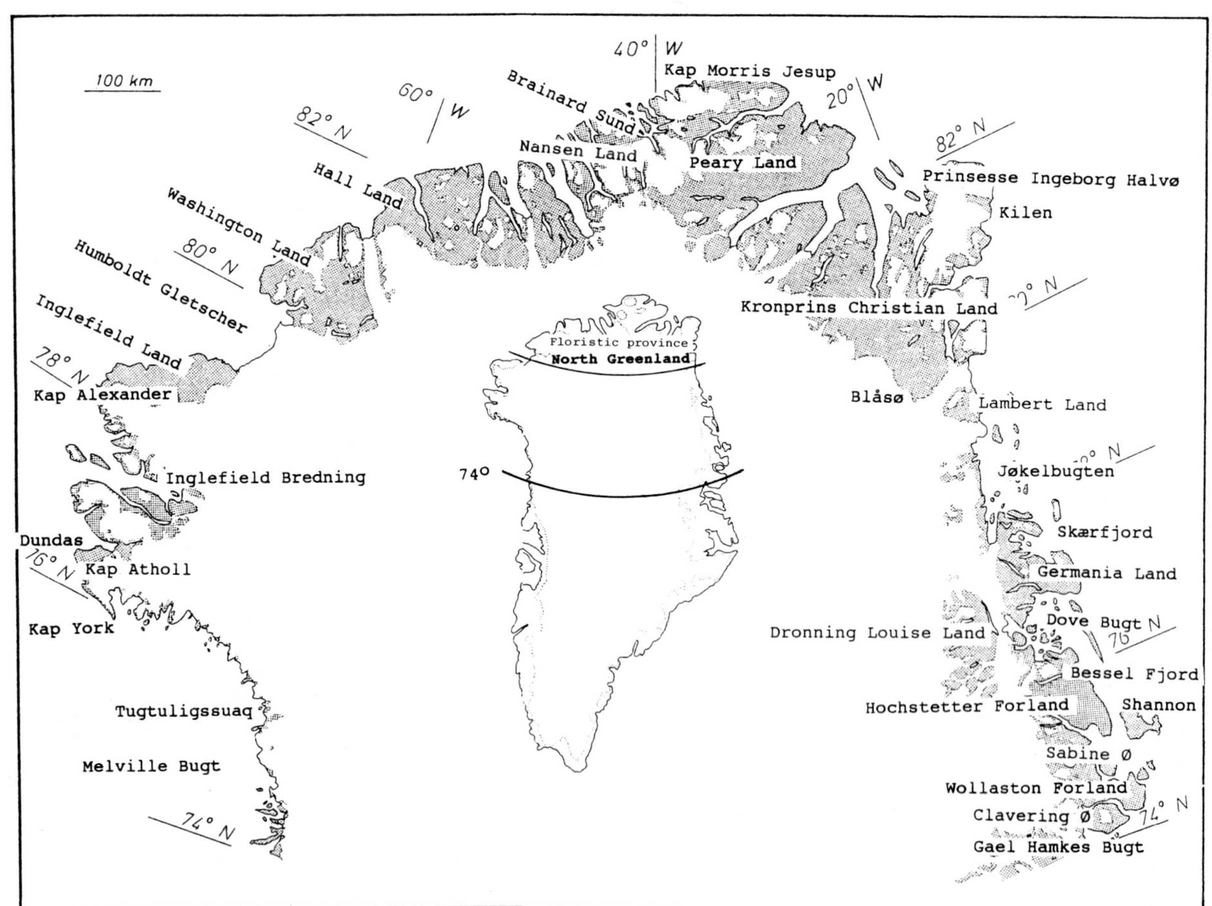

Fig. 1. Map of northern Greenland showing the delimitation of the study area and the floristic province North Greenland. Localities mentioned in the text are shown.

these are herb-slope species reaching the southfacing slopes of Clavering Ø and Wollaston Forland. The corresponding limit for several of these low arctic species in West Greenland has been determined at Upernavik Isstrøm (72°30′N) just south of Melville Bugt (Bay 1983). Generally, most low arctic and high arctic species have their northern and southern limits at c. 70°N, in West and East Greenland respectively. The southern limit at 74°N corresponds to the delimitation of the Canadian High Arctic (Edlund 1983).

2.1. Topography

Most of the land is mountainous, either alpine terrain as in East Greenland or undisturbed plateaus north of 81°N. Huge icecaps and glaciers are widely distributed throughout the area. In North Greenland app. 30% of the area is covered with permanent glaciers. In Melville Bugt the ice-free land is restricted to nunataks, semi-nunataks and islands. Lowlands, i.e. areas below 300 m. a.s.l., are restricted to coastal areas and valleys, covering app. 20% of the ice-free area in North Greenland (Aastrup *et al.* 1986).

The topography of North Greenland north of 79°30′N was influenced by the North Greenland folding resulting in the strongly folded and metamorphosed rocks to be found there today. In the northern part an alpine landscape has been formed, whilst in the southern part horizontal, undisturbed rocks are found, shaped in a plateau landscape reaching 800–1000 m. a.s.l. intersected by valleys and fjords.

The southern part of Northeast Greenland consists of alpine terrain which in the interior parts reaches 1000–2000 m. a.s.l. The area is intersected by fjords almost reaching the Inland Ice. The island Shannon, Hochstetter Forland, and eastern Germania Land are the largest lowlands north of Jameson Land. Numerous islands occur in Dove Bugt and Jøkelbugten north and south of Germania Land, respectively.

2.2. Geology

The oldest bedrock in the area is found at Melville Bugt which exclusively consists of different gneisses (Fig. 2). In addition gneiss is exposed in the areas around the interior parts of Inglefield Bredning and in Inglefield

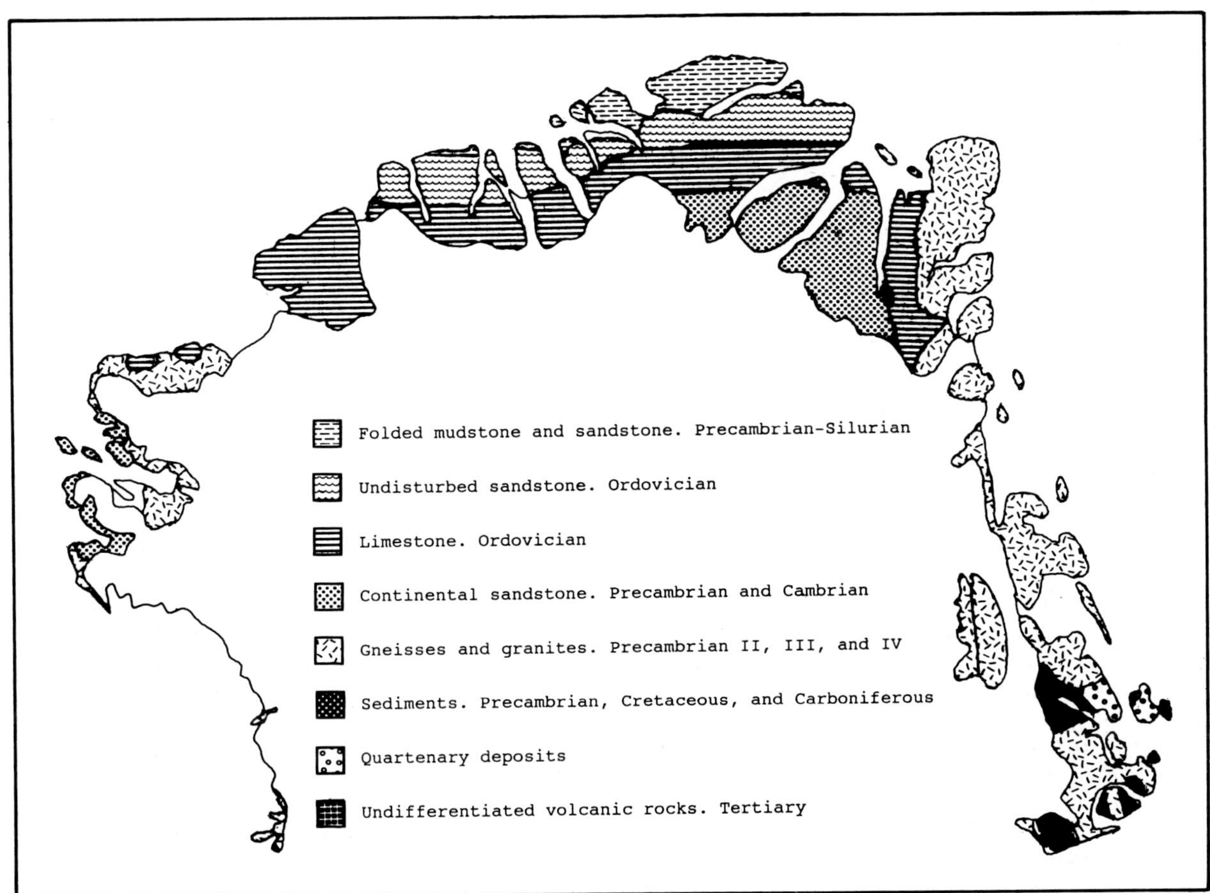

Fig. 2. Geological map of northern Greenland. Type and age are indicated for the dominating rocks. Compiled from Geological map. Greenland Geological Survey, 1984.

Table 1. Climatic data from northern Greenland. Source: Computerized data from the weather stations of Meteorological Institute of Denmark. Temperature is recorded 8 times a day with the exception of Dundas 1968–82 with only 5 records a day. Summer is defined as July and August. 1) Additional data from Mortensen (1987). * Corrected values reduced with 0,4°C due to observations only at 1200 and 1800 GTM. Degree-days reduced with 35 (Qaanaaq) and 24 (Station Nord). ** Estimated from data from Brønlundhus (82°10'N, 30°30'W) (Fristrup 1952).

Site	Recording period temperature/ precipitation	Mean temperature		Precipitation		Degree-days above 0°C
		Annual °C	Summer °C/month	Annual mm	Summer mm, (%)	
Kap Morris Jesup	1981–87/–	−19,0	0,2	–	– –	70,2
83°38'N,		Min: −20,2	−0,2			56,1
33°22'W		Max: −17,1	0,7			87,7
Kap Moltke[1)]	1974–87 ÷83/	−15,7	4,2	<25**	<5** (18)	305,3
82°09'N,	1948–50** –	Min: −16,7	3,5			289,3
29°55'W		Max: −13,8	5,1			319,1
Hall Land	1983–87/–	−20,0	3,3	–	– –	276,1
81°44'N,		Min: −20,3	2,2			206,0
59°00'W		Max: −19,6	4,3			383,8
Station Nord	1961–87 ÷73–74/	−17,1*	2,3*	227,3	48,1 (21)	221,6
81°36'N,	1961–87: ÷72–75; 78–82	Min: −18,4*	0,1*	98,2	15,7	119,6
16°40'W		Max: −15,5*	0,7*	422,7	116,7	338,8
Krøyer Holme	1985–87/–	−12,7	1,1	–	– –	213,2
80°13'N,		Min: −13,3	0,1			111,0
13°43'W		Max: −12,0	2,0			366,6
Qaanaaq	1964–79/1964–77	−10,9*	4,1*	108,2	46,4 (43)	295,7
77°29'N,		Min: −11,0*	3,0*	41,9	2,2	202,9
69°12'W		Max: − 8,9*	4,9*	205,5	79,2	366,4
Danmarkshavn	1961–87 ÷81/do	−12,4	3,0	192,2	44,0 (23)	236,7
76°46'N,		Min: −13,6	2,3	74,4	7,0	171,5
18°46'W		Max: −10,4	4,1	397,2	118,4	346,1
Carey Øer	1982–85 ÷83	− 9,8	2,7	–	– –	–
76°38'N,		Min: −11,2	1,5			
73°00'W		Max: − 8,8	3,3			
Dundas	1961–77 ÷76/do ÷75	−10,8	3,7	173,1	53,4 (31)	292,3
76°34'N,		Min: −12,5	2,1	45,2	4,0	171,3
68°48'W		Max: − 9,2	5,2	458,2	101,5	429,0
Thule Air Base	1974–87/1982–87	−11,6	4,2	163,3	60,7 (37)	348,1
76°31'N,		Min: −13,1	2,8	82,5	82,5	216,6
68°60'W		Max: −10,1	6,1	254,8	94,8	516,8
Tugtuligssuaq	1978–79	− 9,7	4,4	–	—	–
75°21'N,						
58°35'W						
Daneborg	1958–87 ÷75–81/1958–74	− 9,9	3,6	285,5	45,7 (16)	285,2
74°18'N,		Min: −11,3	2,5	129,2	14,7	205,1
20°13'W		Max: − 9,3	4,6	477,8	128,9	376,9
Edderfugle Øer	1982–87	−10,9	3,0	–	– –	238,9
74°02'N,		Min: −13,0	1,5			141,1
57°49'W		Max: − 7,6	4,6			399,6

Land. The main part of the coastal areas in Thule district from Dundas to Kap Alexander is built up of Eocambrium shelf sediments. This formation has a scattered occurrence in Inglefield Land, but is a dominant feature in areas adjacent to the Inland Ice in eastern North Greenland.

The geology of North Greenland is moreover divided into three continuous zones running in west-east direction (Ghisler 1986). A 800 km long formation of limestone deposits dominates in the southern part having the largest extension in Washington Land. Mudstone and sandstone deposits occur north of the limestone zone. After deposition of the whole series of sediments the northern part of the area was influenced by a mountain chain folding.

Northeast Greenland belongs to the Caledonian fold belt. Precambrian basement gneisses are the most widespread formations in the northern part whereas granites and other gneisses are dominant south of 76°N. Undifferentiated volcanic rocks occur in the area Wollaston Forland – the Pendulum Islands – southern Shannon. In Northwest Greenland this formation is only known from the Kap York area. Carboniferous and cretaceous formations are found on the west and east side of Clavering Ø, respectively. Larger quarternary deposits are found at Hochstetter Forland and Shannon. The general features of the geology of the study area are shown in Fig. 2.

2.3. Soils

The soils north of 80°N are termed polar desert soils according to Tedrow (1966) and are within the part of the Arctic with continuous permafrost. Characteristic for their formation in the interior parts are the relatively dry conditions, they are mostly snow-free during winter, and with only a minor component of organic matter resulting in weak horizon development. Commonly, desiccation cracks and salt efflorescences occur in the continental parts. As the soil distribution pattern tends to follow the snow accumulation pattern the exposed snow-free areas south of 80°N have soils belonging to the polar desert, whereas subpolar desert soil and tundra soils are found in depressions and along watercourses.

In only two areas, Tugtuligssuaq, Melville Bugt and Inglefield Land detailed soil investigations have been undertaken (Jakobsen 1988, Tedrow 1970). The soil in the gneissic Melville Bugt is acid with pH ranging between 5 and 6 (Jakobsen 1988). The soils of Peary Land are calcareous, with pH-values between 6 and 8 (Holmen 1957). Acid soils were not found. Gelting (1934) made a number of hydrogen ion determinations in Northeast Greenland. He states that highly calcareous soil is commonly distributed within the area (south of 75°N), the sediments often consisting of lime. Highly acid soil hardly occurs whereas slightly acid soil sometimes occurs in the gneissic and granitic areas, but there, too, calcareous soil is locally present. The raised marine terraces sometimes contain a considerable amount of lime originating from the shells embedded there.

2.4. Climate

The climate of northern Greenland is high arctic characterized by low temperatures, low precipitation, and only 2–3 months with mean temperature above 0°C.

Systematic meteorological data on temperature and precipitation from the 12 permanent weather stations have been compiled by the Danish Meteorological Institute and Mortensen (1987) (Table 1, Figs 3 and 4). Information from the temporary station at Tugtuligssuaq, Melville Bugt (Jakobsen *et al.* 1980) is included

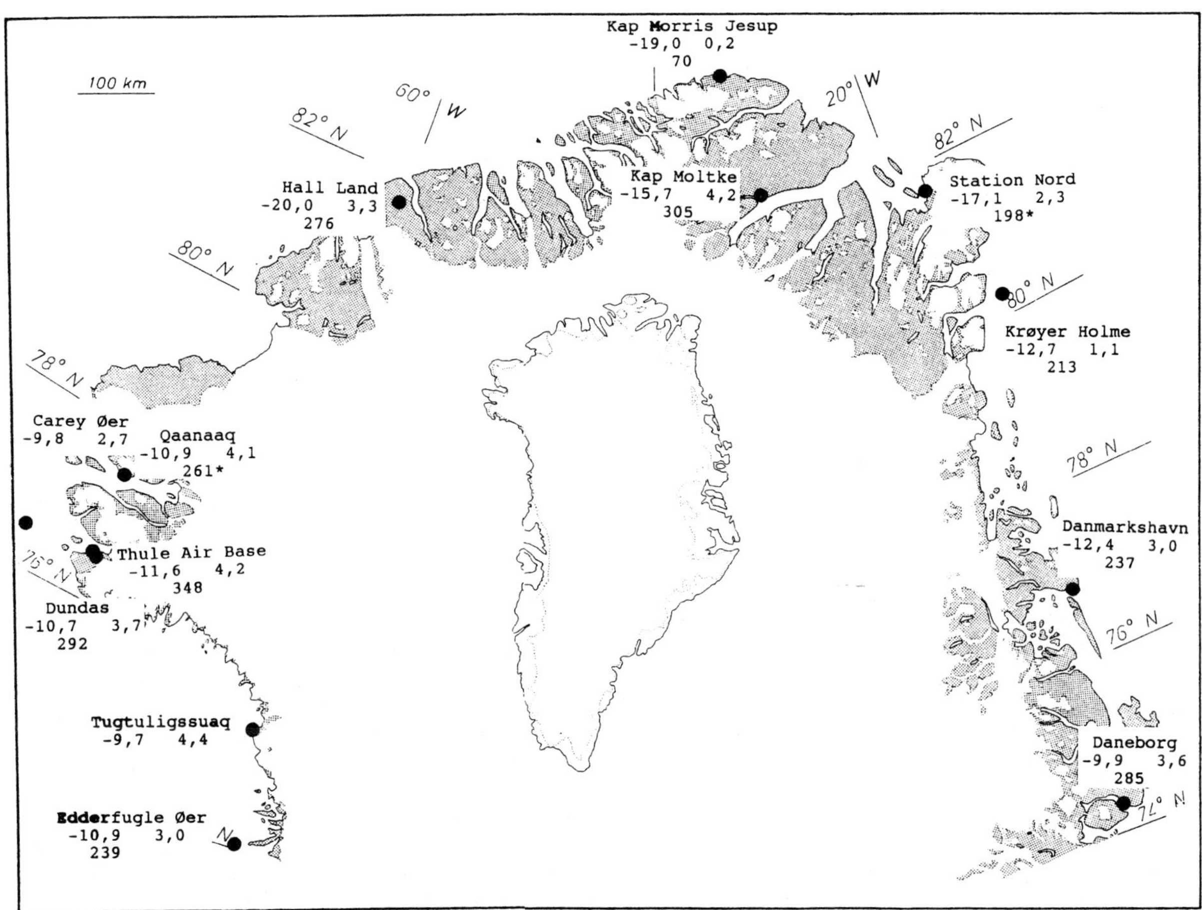

Fig. 3. Annual mean temperature (°C), summer (July-August) mean temperature, and sum of degree-days above 0°C (lower number) in northern Greenland. Recording periods are given in Table 1.

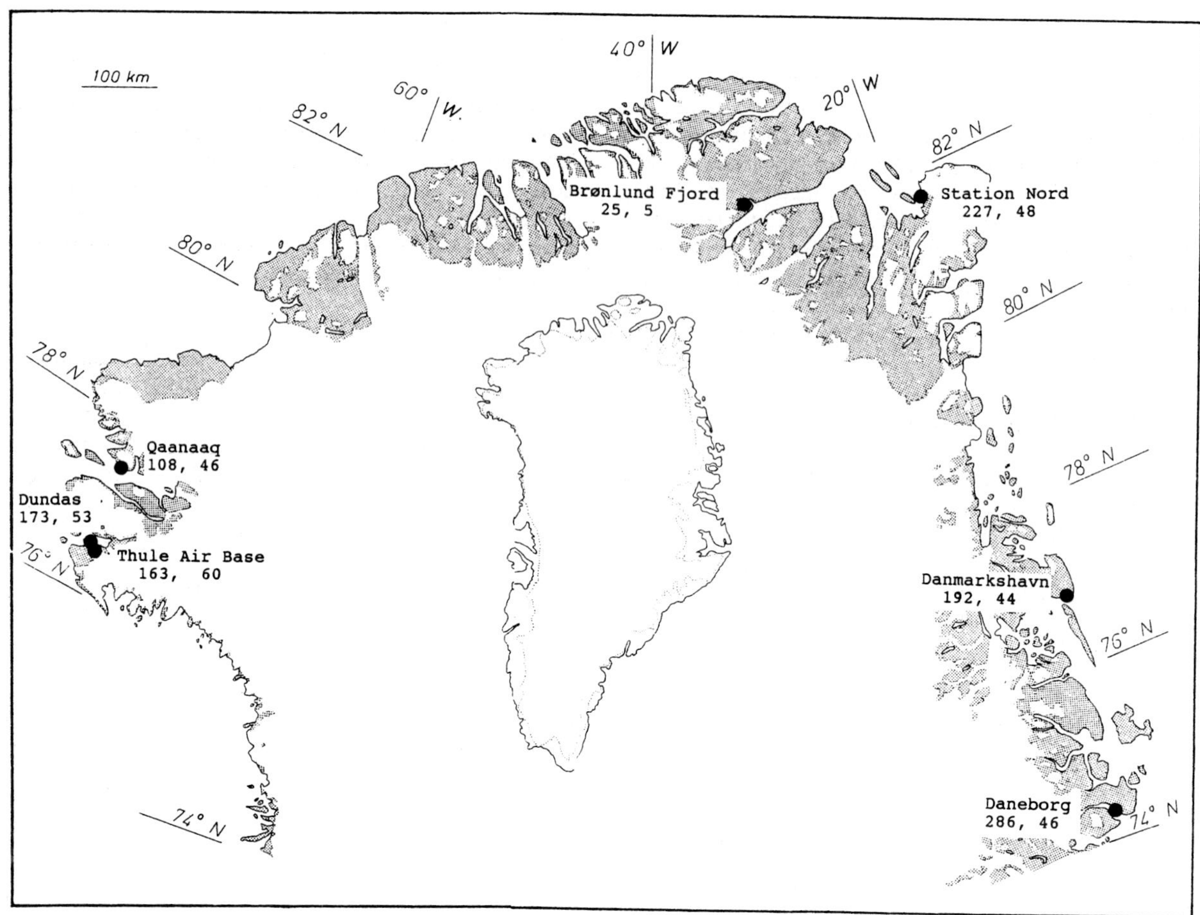

Fig. 4. Annual mean precipitation (mm), and summer (July-August) mean precipitation in northern Greenland. Recording periods are given in Table 1.

too. Only five stations have data from more than 25 years. All stations north of 79°N except for Kap Moltke and Station Nord started the recordings in the 1980's, consequently the available data are limited. The data are not directly comparable to those originating from different recording periods.

2.4.1. Temperature

For each station the monthly and yearly mean temperatures have been calculated for the recording periods. The summer and yearly mean temperature are given in Table 1. Summer is defined as the period from July 1 to August 31. Temperature is mostly recorded 8 times a day.

The monthly means from Station Nord and Qaanaaq have been reduced with 0,4°C due to only two daily observations at 1200 and 1800 GMT giving a too high mean daily temperature (Cappelen pers. comm.). Consequently the degree-days from these stations are too high and corrected with 0,4°C time the number of days with positive mean temperatures. The mean annual temperature varies from app. −10°C in the southern part of the area to app. −20°C in the northern part. The summer temperature, which is of main biological importance, varies from app. 3–4°C in the southern parts to app. 0–4°C in the northern. A comparison between the summer mean temperature at the only inland station in North Greenland, Kap Moltke, and the outer coast stations, shows the significant difference between the cold outer coast and the much warmer inland. The most extreme temperature differences between mid winter and summer of 60°C were observed in 1979 at Kap Moltke in Peary Land (Mortensen 1987).

The degree-days are given as the sum of mean daily temperature above 0°C per year. Most of the outer coast stations have values between 200 and 300 with maximum values at Daneborg (285) in the southern part and minimum values of 70 at Kap Morris Jesup. Kap Moltke differs from most of the other stations in this respect, too, as the number of degree-days exceeds 300.

2.4.2. Precipitation

The monthly and yearly precipitation have been recorded at only seven stations which were manned in periods of between 13 and 29 years. The summer and yearly mean precipitation are given in Fig. 4 and Table 1. The annual precipitation is low, ranging from 285 mm in the southeastern part to c. 25 mm in the inland at the northernmost station (Fristrup 1952). 20–30% of the annual precipitation fall as rain during the short summer, ranging from 16 to 43%. Generally, the annual precipitation decreases with latitude, the only exception being Station Nord. The east coast has higher precipitation than the west coast in northern Greenland.

The precipitation is unequally distributed in the landscape during winter because shortly after it is falling most of the snow is redistributed by the wind and accumulated in the leeside of slopes and hills. Thus, the only available precipitation in the snow-free areas is between 5 and 60 mm yearly. The available precipitation in snow-covered areas is further reduced as part of the snow evaporates. It should be noted that the amount of water available to the plants increases during summer when the active layer thaws.

3. Exploration of the study area

3.1. History of botanical exploration

A total of 122 collectors has contributed with information about the distribution of vascular plants in the study area.

A list of botanical expeditions and other expeditions which in one way or another have obtained botanical information from northern Greenland is given in Table 2.

The exploration of the flora of North Greenland started very late compared to that of other parts of Greenland due to the remoteness of the area. Until 1917 only very few collections of plants had been brought home from North Greenland (Bessels 1879, Hart 1880, Greely 1886, Simmons 1909, Ostenfeld & Lundager 1910). During "The First Thule Expedition" in 1912 P. Freuchen collected plants in Peary Land and Danmark Fjorden (Ostenfeld 1915).

Thorild Wulff was the first botanist who made systematical collections of plants and described the vegetation during "The Second Thule Expedition" in 1916–18. His work was published after his death by Ostenfeld (1919, 1923a–b and 1925a–c). By the establishment of the "Danish Peary Land Expeditions" in 1947 the botanical exploration of the eastern North Greenland was initiated. K. Holmen who parcitipated in 1947–50 collected a huge material of vascular plants and mosses and made descriptions of the vegetation (Holmen 1955, 1957). Additional floristic work was carried out by Fränkl (1955) and by B. Fredskild on "The Second Peary Land Expedition" in 1963 (Fredskild 1966a).

In the years 1977–86 geologists collected material from most parts of North Greenland; from Washington Land in west to Kronprins Christian Land in east in connection with their geological mapping projects. The largest collections were made by Ole Bennike and Svend Funder. In 1985 the Greenland Fisheries and Environmental Research Institute made a biological reconnaissance in North Greenland from Hall Land to Kronprins Christian Land, in which the present author participated (Aastrup *et al.* 1986, Fredskild *et al.* 1986). Additional collections were brought home in connection with botanical projects in Kronprins Christian Land (Fredskild *et al.* 1987, Fredskild & Bay 1988), Peary Land (Fredskild & Bay 1988), and Nansen Land (Fredskild & Bay 1992).

The botanical exploration of Northwest Greenland from 74°–80°N began much earlier. John Ross was the first to bring plants home in 1818. His plant lists, however, are of minor importance as there is much uncertainty about his localities as he collected on both the west coast of Greenland and the east coast of North America. The most doubtful collections from the 19th century are discussed by Simmons (1904). Several collectors followed during the 19th century and characteristic to these are that they contributed with only a small number of species from a few localities. Simmons was the only one in the last century who made systematical collections in the Thule district. In his paper (Simmons 1909) he gives a comprehensive discussion of these early expeditions to Northwest Greenland.

As a result of the establisment of the Thule Station (Dundas) by Knud Rasmussen in 1910 the botanical exploration of Northwest and North Greenland was intensified. Ostenfeld (1905, 1923b, 1923c, 1923d, 1925c) has published material from several collectors from Thule district and North Greenland. The most extensive work was done in Inglefield Land by Noe-Nygaard during "The Danish Jubilee Expedition north of Greenland 1920–22", when 97 of the 150 species of vascular plants known from Thule district were found. In 1991 Inglefield Land was visited by Nielsen and Ulbæk who brought home collections.

Thorlaksson has contributed with a large collection from Dundas collected during "The Danish Thule and Ellesmere Land Expedition" in 1939–40. Thule district has been visited by a number of naturalists resulting in a great number of collections from various localities in the district. The following important collectors should be mentioned: Fristrup, Just, Tang Lassen, Higgs, Lang and de Bonneval. As they are "non-professional botanists" some plant groups are underrecorded, especially grasses, sedges, and inconspicuous species as *Sagina* spp., *Draba* spp., and *Koenigia islandica*. Systematical collecting has been carried out at five localities in the district by Greenland Botanical Survey in 1977 and 1980

Table 2. List of botanical expeditions and other expeditions which have contributed with larger collections. The collectors and the references are given.

Year	Expedition, institution/Collector	District	Publication
1871–72	Die Amerikanische Nordpol-Expedition	N	Bessels 1879
1875–76	British Polar Expedition of 1875–76/H. Hart Coppinger	N	Hart 1880
1881–84	Lady Franklin Bay Ekspedition	N	Greely 1886
1886–87	Ryders Ekspedition/Ussing	NW	Lange 1887
1898–02	The Second Arctic Expedition in the "Fram"/Simmons	NW	Simmons 1909
1906–08	Danmark-Ekspeditionen/A. Lundager, I. P. Koch	NE, N	Ostenfeld & Lundager 1910
1912	The First Thule Exp./P. Freuchen	N	Ostenfeld 1915
1914–16	Crockerland Expedition/W. E. Ekblaw	NW	Ekblaw 1919
1916–18	The Second Thule Exp./T. Wulff	N, NW	Ostenfeld 1923b
1921	The Danish Jubilee Exp. north of Greenland 1920–22/J. Noe-Nygaard	NW	Ostenfeld 1925c
1928	Godthåb Exp. 1928/G. Seidenfaden	NW	Seidenfaden 1932
1929–30	The Norwegian Scientific Expedition to East Greenland	NE	Vaage 1932
1931–34	Three Year Expedition 1931–34 to Northeast Greenland /G. Seidenfaden, T. Sørensen	NE	Gelting 1934, 1937, Seidenfaden & Sørensen 1933, 1937, Sørensen 1933, 1945
1936	Den Naturhistoriske Ekspedition til Nordvestgrønland 1936 /F. Salomonsen	NW	Sørensen 1943
1937	The Louise A. Boyd Arctic Expedition of 1937 and 1938/H. Oosting	NE	Oosting 1948
1938–39	Dansk Nordøstgrønlands Ekspedition 1938–39/P. Gelting	NE	
1939–40	Den Danske Thule og Ellesmere Land Ekspedition 1939–40/G. Thorlaksson, C. Vibe		
1947–50	Dansk Peary Land Ekspedition 1947–50/K. Holmen	N	Holmen 1957
1950	Greenland Geological Survey mapping project 1950/K. Jakobsen	NW	
1958	Operation Hazen/J. Powell	N	Powell 1971
1963	2. Peary Land Expedition 1963/B. Fredskild	N	Fredskild 1966a, 1966b
1976	Joint Services Expedition to North Peary Land 1969/A. Griffin	N	Griffin 1972
1976	Dansk-Svensk Ekspedition til Nordøstgrønland/H. Bruck, J. Mikaelsson, L. Adrielsson, M. Elander	NE	
1977–84	GGU geological mapping activity/O. Bennike, S. Funder	N	
1977	Greenland Botanical Survey 1977/B. Fredskild	NW	Fredskild *et al.* 1978
1979	Kund Rasmussen Memorial Expedition 1979/C. Bay & B. Fredskild	NW	Fredskild *et al.* 1979
1980	Knud Rasmussen Memorial Expedition 1980/B. Fredskild	NW	Fredskild 1985, Fredskild & Møller 1981
1980	British North-east Greenland Expedition 1980/G. Halliday	NE	Halliday 1981
1985	Greenland Fisheries and Environmental Research Institute/C. Bay	N	Aastrup *et al.* 1986, Fredskild *et al.* 1986
1987	Irish Expedition to North-east Greenland 1987/R. Goodwillie	NE	Cabot *et al.* 1988
1988	Greenland Botanical Survey 1988/C. Bay	NW	Fredskild & Bay 1989
1988–90	Greenland Home Rule Authorities mapping project 1988–90 /C. Bay, B. Fredskild	N	Bay & Boertmann 1989, Bay & Fredskild 1990, 1991
1990	British North-East Greenland Expedition 1990/G. Halliday, R. Corner, H. Lang, R. & S. David, J. Hooson, T. Meiklejohn, C. Wells	NE	Halliday & Corner 1991
1991	Herbivores in the High Arctic/C. Bay	N	Fredskild *et al.* 1992 Klein & Bay in prep.

(Fredskild *et al.* 1978, Fredskild & Møller 1981) and by the present author in 1988 (Fredskild & Bay 1989).

Sørensen (1943) summarizes the botanical exploration of the Melville Bugt in the southwestern part of the study area. His paper is mostly based on the collections by Salomonsen who participated in "The Natural History Expedition to Northwest Greenland" in 1936. Additional investigations have been carried out by Jakobsen (1950) in 1950 and by Fredskild and Bay in 1979 during "The Knud Rasmussen Memorial Expedition, 1979" in the northern part of Melville Bugt (Fredskild & Bay 1980).

The botanical exploration of Northeast Greenland was initiated in 1823 when Sabine brought home a number of vascular plants from Sabine Ø (Hooker 1925). "Der Zweite Deutsche Nordpolexpedition 1869–70" contributed with the first large collection of plants from Northeast Greenland, mainly from the coast of Wollaston Forland and adjacent islands (Buchenau & Focke 1874). But it was not until the turn of the century that Danish botanists came to the area. Hartz and Kruuse (Kruuse 1905) visited a few localities in the present study area during the expedition to the east coast in 1900.

The Clavering Ø-Wallaston Forland area (74°–75°N) on the east coast is the most intensively investigated part of the study area. Seidenfaden, Gelting, and Sørensen visited it during the "The triennial Danish East-Greenland-Expedition 1931–33" and have published a number of papers (Gelting 1934, 1937, Seidenfaden &

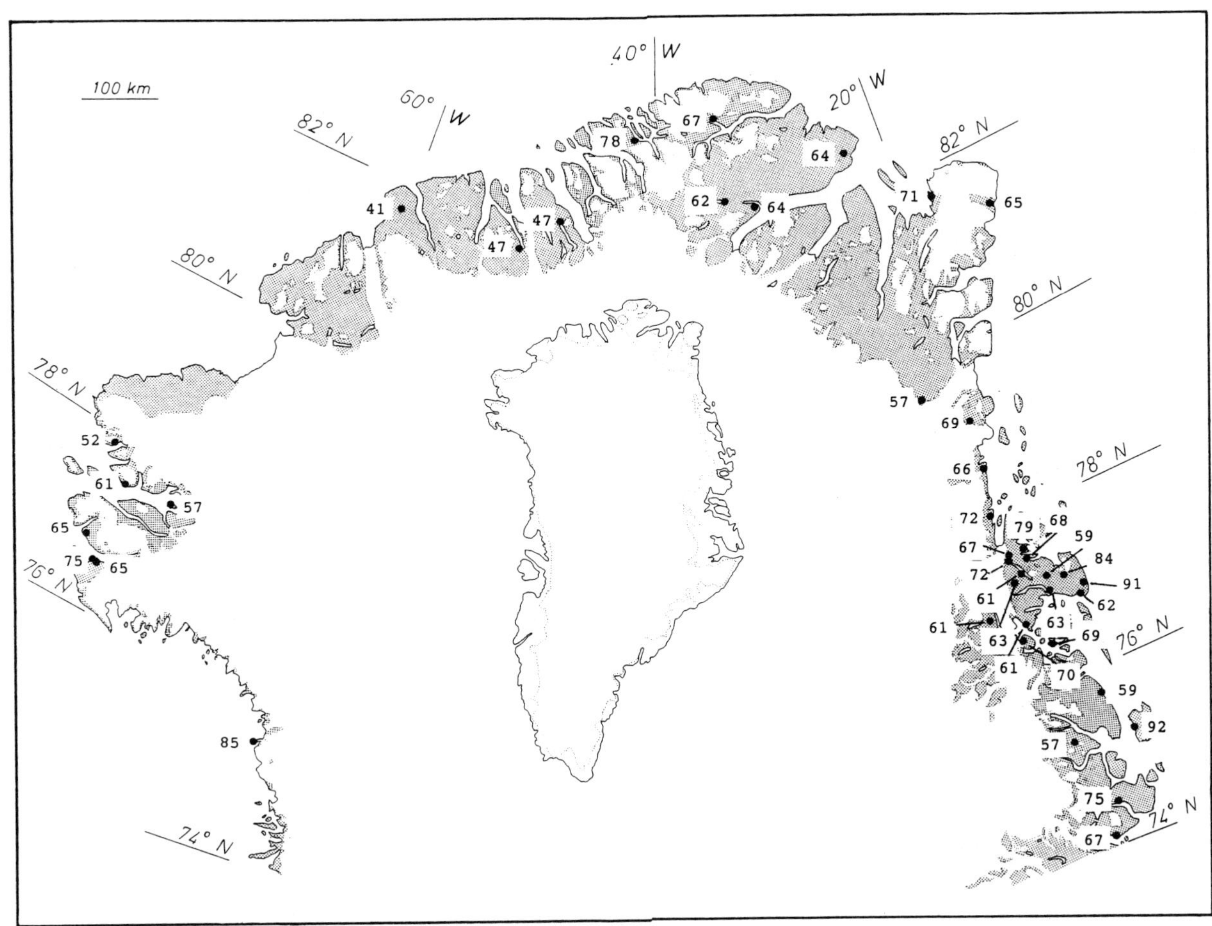

Fig. 5. Number of taxa from localities considered well-investigated as percentages of the total number of taxa in the districts. District borders are indicated in Fig. 8.

Sørensen 1933, 1937, Sørensen 1933, 1935, 1941, 1945). In the same years Norwegian botanists visited the same area and Vaage (1932) published some of the floristic results. Schwarzenbach (1961) deals with the nunatak zone in the southern area. The "Dansk-Svensk Ekspedition til Nordøstgrønland" in 1976 and "British North-East Greenland Expedition 1980" under the leadership of Halliday have contributed with valuable material from the southern area (Halliday 1981). During "British North-East Greenland Expedition 1990" to Kuhn Ø and Th. Thomsens Land (74°40′–75°10′N) Halliday and other British botanists carried out floristic work and found three species new to the study area (Halliday & Corner 1990).

Stemmerik and Hauge Andersson brought home large collections in 1987–88 in connection with their geological and geodetical work in the area between Clavering Ø and Bessel Fjord.

The northern area south of Skærfjord (75°–78°N) has been treated by Ostenfeld & Lundager (1910), Lundager (1912), and Seidenfaden & Sørensen (1937). Gelting carried out botanical investigations during "Dansk Nordøstgrønlands Ekspedition 1938–39" on the south coast of Germania Land. All the available material from this expedition is published in the present paper.

Half a century passed before the botanical investigations in Northeast Greenland were continued when "The Irish Expedition to North-east Greenland 1987" visited hitherto unknown parts of northern Germania Land (Cabot *et al.* 1988). Goodwillie carried out the botanical work on the expedition. The Greenland Home Rule Authorities pointed out the area between 75°N and Lambert Land (79°30′N) as study area for a three year biological-archaeological mapping project in 1988–90, in which Fredskild and the present author participated (Bay & Fredskild 1990, 1991).

Detailed investigations by Mølgaard and Bay have been carried out in minor areas (79°40′–81°40′N) in Kronprins Christian Land in Northeast Greenland: Kilen, Prinsesse Ingeborg Halvø, and Blåsø area (Mølgaard 1985, Fredskild *et. al.* 1987, Klein & Bay 1990).

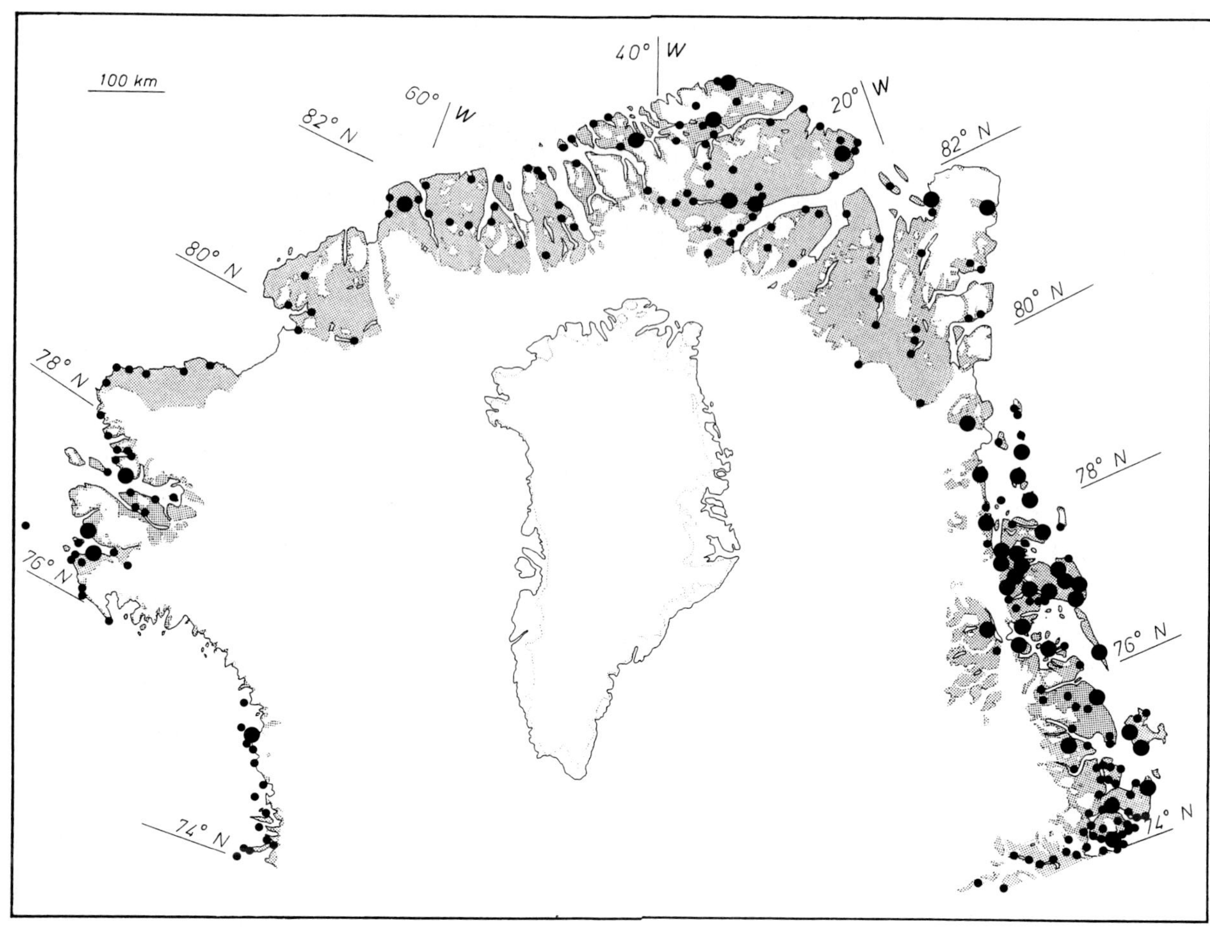

Fig. 6. Botanical activity of the study area. Localities with more than 60% of the potential species in the district (●), and localities with 10–60% of the species (•) are mapped. Half of the total number of localities has less than 10% of the species in the district and are omitted. Further explanation is given in the text.

3.2. Investigation activity

A prerequisite to draw reliable phytogeographical conclusions from distribution maps is a comparison between the dot maps and a map showing the intensity of botanical activity, i.e. number of species recorded at the localities compared to the total number known from the respective districts. As the number of taxa decreases with increasing latitude and oceanity (Fig. 8) the number of taxa from localities considered well-investigated is compared to the number of taxa in the districts (Fig. 5). These are defined as localities that have been visited by trained botanists for at least 3 days. Records kept during the field work in 1989 revealed that more than 95% of the species from a locality investigated for more than 6 days was found during the first 3 days (Bay & Fredskild 1990). Generally, 60%–80% of the total number of taxa known from a district is present at well-investigated localities. The exception is three localities considered well-investigated situated in the interior North Greenland that are under this limit of 60%. The percentages of the three well-investigated outercoast localities in Northeast Greenland have been altered as few occurrences of a species outside its main distribution are accepted. Thus, the very few finds of species that are almost restricted to the inland have been neglected when calculating the percentages for the localities. The localities are classified in three categories according to the number of species present as a percent of the total number of species in the district (Table 3). Well-investigated (> 60%) and less intensively investigated (11–60%) localities are mapped in Fig. 6.

In half of the 707 localities only few species (1–20) have been collected and only four localities have more than 80% of the potential species in the district. Tugtuligssuaq is the only one at the west coast and three of these are investigated by the author.

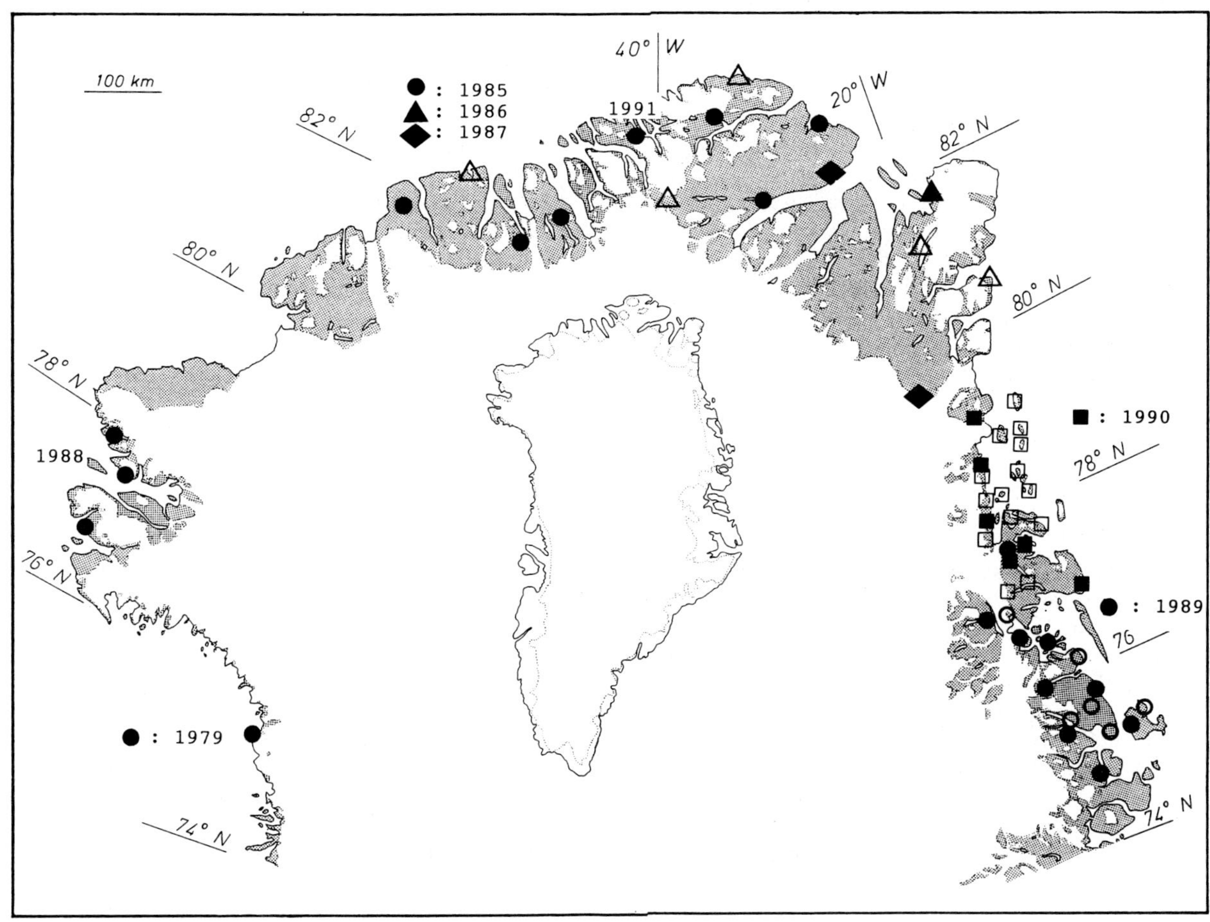

Fig. 7. Localities investigated by the present author. Closed symbols indicate well-investigated localities. Open symbols indicate localities investigated during few hours.

4. Materials and methods

4.1. Field work

Field work was carried out by the author in 1979 and 1985–91. North, Northwest, and Northeast Greenland were investigated in 1985–87 plus 1991, 1988, and 1989–90, respectively. The localities are shown in Fig. 7. The field work in 1985 was part of an environmental reconnaissance by the Greenland Fisheries and Environmental Research Institute, Copenhagen, whereas the work in 1986 and 1988 was primarily a phytogeographical field work financially supported by the Commission of Scientific Research in Greenland. The 1989 and 1990 field work was part of a biological mapping project financially supported by the Greenland Home Rule Authorities from a grant donated by "Aage V. Jensen Charity Foundation". In connection with wildlife

Table 3. Number of species in the districts in each of the three categories. Localities in category 3 are considered well-investigated.

Category		District							
		1	2	3	4	5	6	7	8
1	0–10%	1–10	1–15	1–6	1–12	1–6	1–15	1–7	1–19
2	11–60%	11–59	16–90	7–33	13–72	7–33	16–87	8–43	20–115
3	61–	60–	91–	34–	73–	34–	88–	44–	116–
Nos of species		98	150	55	120	55	145	72	191

projects material was collected in Peary Land and Nansen Land in 1987 and 1991, respectively. In addition material collected during the field work in Melville Bugt in 1979 has been included.

Specimens of all vascular plant species were collected at each camp site on areas covering up to 100 km^2. The duration of the stay at each site mostly was three to seven days. However, the visits in 1986 lasted a few hours whereas they lasted 3–4 weeks in 1987 and 91. At each locality all types of landscape were searched in order to collect plants from all vegetation types present.

Geologists, geodesists, and English biological expeditions were encouraged to collect in areas not visited by the author. A total of c. 10,500 collections has been brought to the Greenland herbarium of Botanical Museum, Copenhagen during this period.

4.2. Herbarium study

More than 24,700 specimens from northern Greenland, kept in C, were examined and additional material from O, CAN, and LAN was included. After determination the specimens were listed and mapped. The revision of the material has resulted in many redeterminations and four taxa new to Greenland have been recognized.

Collections from a number of older expeditions are published in the present paper. These include "Dansk Nordøstgrønlands Ekspedition 1938–39", "Den Danske Thule og Ellesmere Land Ekspedition 1939–40", and "Dansk-Svensk Ekspedition til Nordøstgrønland" in 1976.

4.3. Taxonomical considerations

Generally, the taxonomy and nomenclature follow Böcher *et al.* 1978. Any deviation from this is mentioned. The main diagnostic characters for delimiting closely related taxa are discussed along with notes concerning taxonomy and distribution of the species.

In the following, all exceptions are mentioned. In some cases closely related taxa have been treated as one taxon if the distinction between them seems to break down in the North Greenland material. The nomenclature of the new taxa to Greenland follows Hultén (1968): *Pedicularis sudetica* Wild. ssp. *albolabiata* Hult. and *Geum rossii* (R. Br) Ser., Porsild (1964): *Puccinellia bruggemanni* Th. Sør., and Löve & Löve (1961): *Phippsia algida* (Sol.) R. Br. ssp. *algidiformis* (H. Sm.) L. & L.

Information from literature on the distribution of species outside the range recorded from the material studied is annotated.

The taxa are arranged alphabetically.

Antennaria compacta Malte
Porsild (1965) has revised the arctic material of *Antennaria*. The *A. compacta* material, determined by Porsild and kept in O and CAN, has been examined. This material is not convincingly different from *A. ekmaniana* A. E. Pors. and has been redetermined as this species. True *A. compacta* is considered not occurring in the study area.

Arenaria humifusa Wbg.
Halliday (1981) has recorded the species as far north as Hochstetter Forland (75°19′N). As the specimen has not been available it is omitted on the map.

Betula nana L.
Eastwood (1984) records *Betula nana* from Germania Land. Having worked in this area in Northeast Greenland in 1989 it seems to me unlikely that *Betula nana* should have been collected in the high arctic Germania Land (77°–78°N).

Braya intermedia Th. Sør.
Sørensen (1954) published this as a new species of *Braya*. It is intermediate between *B. linearis* Rouy and *B. humilis* (C. A. Mey.) Robins. He proposed it an amphidiploid hybrid often with a large number of undeveloped ovules. The material from the study area is not convincingly different from *B. humilis* and consequently, the species is included in *B. humilis*. Recently J. G. Harris (in sched.) came to the same conclusion when revising the Greenland material.

Cerastium alpinum L. ssp. *lanatum* (Lam.) Asch. & Graebn.
See *C. arcticum*.

Cerastium arcticum Lge. *s.l.*
The material of the *Cerastium alpinum/arcticum* complex has been treated as one taxon: *C. arcticum* Lge. as no convincing *C. alpinum* was found. The material from district North Greenland is very uniform whereas some of the collections in the southern parts seem to be more related to *C. alpinum*.

Cerastium regelii Ostf. ssp. *caespitosum* (Malmgr.) Tolm.
According to Heide *et al.* (1990) the closely related species described by Hultén (1956) under the name *C. jenisejense* should be included in *C. regelii* giving it a circumpolar distribution.

Cystopteris fragilis (L.) Bernh. *s.l.*
The material has not been separated in the two subspecies: ssp. *dickieana* (Sim) Hyl. and ssp. *fragilis* according to Böcher *et al.* (1978) because only a minor part of the material was fertile.

Draba adamsii Led.
Draba micropetala Hook. can not be distinguished from *D. adamsii* when flowers are absent. Consequently, it is

included in *D. adamsii* as the major part of the material was in fruit.

Draba alpina-complex
The complex includes three species: *D. alpina* L., *D. bellii* Holm, and *D. gredinii* Ekm. *D. gredinii* is included in *D. alpina,* see below.

The pubescence on the basal leaves of *Draba alpina* is composed of simple, forked, and branched hairs. These hairs are often concentrated on the margin of the leaves. The stem is less hairy than *D. bellii*. The shape of the siliques is ellipsoid and acuminated and mostly glabrous or with few hairs.

D. bellii has more narrow basal leaves than *D. alpina* with a much denser layer of usually branched hairs. The uppermost 1/3 of the stem is very hairy. Small unbranched hairs give the siliques a hairy appearance. These hairs are not developed on unripe siliques.

Draba arctica J. Vahl
See *D. cinerea*-complex.

Draba bellii Holm
See *D. alpina*-complex.

Draba cinerea-complex
The North Greenland collections with stellate hairs on leaves as well as on siliques are considered one taxon: *Draba arctica* J. Vahl. It has not been possible to make a convincing distinction into more taxa as done by Böcher (1966). Much of the material formerly considered *D. cinerea* because of small siliques and a low growth seems to be hybrids as the siliques are twisted and not well-developed. A few collections from Northwest Greenland may prove to be true *D. cinerea,* but are mapped here as *D. arctica.*

Draba gredinii Ekm.
This taxon is included in *D. alpina* L. as it is impossible to determine the colour of the petals on dried material. The hairs on the leaves of specimens determined by Ekmann (1933) as *D. gredinii* very much resemble those on typical *D. alpina* from Northeast Greenland. Consequently the material is treated as one taxon.

Draba macrocarpa Adams
D. macrocarpa Adams is included in *D. bellii* Holm.

Draba micropetala Hook.
D. micropetala is included in *D. adamsii* Led.

Draba nivalis Liljebl.
The only two collections from North Greenland determined by Holmen are rejected as the material is too poor for a safe determination.

Draba norvegica Gunn.
No true *Draba norvegica* has been examined. A few poorly developed collections may be *D. norvegica* but are included in *Draba glabella* Pursh.

Draba oblongata R. Br.
D. oblongata is synonymous with *D. adamsii* Led.

Draba subcapitata Simm.
Collections of this species show some variation in the hairiness. They are often very hairy on the stem especially on the uppermost 1/3, whereas some specimens only have few hairs and then only on the lower part.

Dryas
The western *Dryas integrifolia* M. Vahl and the eastern *D. octopetala* L. have a common area in North and Northeast Greenland, and hybridize in this area (Elkington 1965). Apart from unquestionable collections of these two taxa quite many collections are considered hybrids. The two diagnostic characters used in distinguishing between the species are: 1) occurrence of "octopetala scales" (Elkington 1965) and 2) crenation of the leaves.

D. octopetala is defined by having typical "octopetala scales" and being crenated to the tip of the leaves.

D. integrifolia is defined by lacking scales and lacking crenations or having only 1–2 at the basal part of the leaves.

Other combinations of the characters are considered hybrids. The most widespread hybrid is one with leaves crenated to the tip but without "octopetala scales". The hybrids are very common in eastern North Greenland and Northeast Greenland. A general trend in the North Greenland material is that the influence of *D. octopetala* is reduced westwards.

Dupontia fisheri R. Br./*D. psilosantha* Rupr.
The genus *Dupontia* is represented in Greenland by two taxa: *Dupontia fisheri* R. Br. and *D. psilosantha* Rupr. (Böcher *et al.* 1978). Hultén (1962, 1968), Tsevelev (1983), Porsild (1964), Porsild & Cody (1979), and Scoggan (1978) consider these species as subspecies of *D. fisheri: Dupontia fisheri* R. Br. ssp. *fisheri* and *D. fisheri* R. Br. ssp. *psilosantha* (Rupr.) Hult.

The distinguishing characters used in these papers are: 1) shape and length of glumes, 2) the hairiness of the lemmas, 3) the length of anthers, 4) the shape of the inflorescence, and 5) presence of awn. When comparing the floras it is evident that there is some confusion about the length of anthers and the presence of awn in the two taxa. Böcher *et al.* (1978) and Hultén (1964, 1968) state that *D. psilosantha* has the longest anthers, whereas Scoggan (1979) and Porsild (1964) state that the anthers of *D. psilosantha* are "distinctly shorter than *D. fisheri*". There are no significant differences in length of anthers in the present material determined by the author. *D. fisheri* varies between 1,6 mm and 2,5 mm with most specimens about 2,0–2,4 mm, whereas *D. psilosantha* varies between 1,6 mm and 2,8 mm, mostly within a length of 2,0–2,5 mm. All the floras except Böcher *et al.* (1978) state that *D. fisheri* has an awn in contrast to *D. psilosantha.* Examination of the collections revealed that 66% of the *D. fisheri* material has an awn and 36% of *D. psilosantha* are awned. Consequently, these characters are neglected and the shape and length of glumes in addition to the shape of the inflorescence are used as the main diagnostic characters. The glumes of *D. fisheri* vary between 3,5 and 4,5 mm whereas *D. psilosantha* has more acute glumes of 5–8 mm. The inflorescence of *D. fisheri* is compressed in contrast to the extended inflorescence of *D. psilosantha.* 20% of the typical *D. psilosantha* had hairy glumes.

Honckenya peploides (L.) Ehrh. var. *diffusa* (Horn.) Mattf.
Simmons (1909) has recorded this species from Foulke Fjord, Inglefield Land, which is the northernmost record in West Greenland. As the specimen has not been available this information was omitted on the map.

Papaver radicatum Rottb. *s.l.*
Böcher *et al.* (1978) state that three species are recognizable in the North Greenland part of the distribution of *P. radicatum* coll., namely *P. radicatum s.s.* and two undescribed entities, one with $2n = 84$ and the other with $2n = 70$. Since it was not possible to recognize the two undescribed entities on the basis of herbarium material the taxon is treated here as a single species.

Pedicularis sudetica Willd. ssp. *albolabiata* Hult.
The taxomony follows Hultén 1968.

Phippsia algida (Sol.) R. Br.
All the Greenland material has been examined and separated into two taxa: *Phippsia algida* (Sol.) R. Br. and *P. algida* (Sol.) R. Br. ssp. *algidiformis* (H. Sm.) L. & L. although Mosquin & Hayley (1966) state that *Phippsia* is a monotypic genus. The hairiness, shape of the lemmas and colour of the nerves are used for the separation of the material into the two taxa in addition to the shape of the fruits. *P. algida* ssp. *algidiformis* is characterized by hairy lemmas with a length of 1,5–2,3 mm and with an acute apex of the fruit whereas *P. algida s.str.* has glabrous, light green lemmas of 1,0–1,5 mm and with a rounded apex of the fruit. Moreover *P. algida* ssp. *algidiformis* has distinct lateral, purple nerves and longer erect stems and more extent inflorescence. On the other hand, the main part of the North Greenland material north of 80°N differs from the typical *P. algida s.str.* in the above mentioned characters of the lemmas and is consequently considered a subspecies, that is stated to be a fertile hybrid (Tsvelev 1983). The original nomenclature by Smith (1914) was changed by Löve & Löve (1961) from *Catabrosa concinna* ssp. *algidiformis* to *Phippsia algida* ssp. *algidiformis.* This subspecies is a new taxon to the flora of Greenland and its distribution is high arctic. It is found north of Scoresby Sund through North Greenland to 76°N on the west coast.

Poa alpina L. var. *vivipara* L.
The material from Ingolf Fjord, 80°36′N, mentioned in Böcher *et al.* (1978) is rejected as inadequate for a positive determination.

Poa hartzii Gandoger *s.l.*
The viviparous form *forma prolifera* (Simm.) Boivin is included in the taxon. It has only been found twice in the eastern North Greenland (Bay in prep.).

Potentilla hookeriana Lehm. *s.l.*
All the material, except for 9 collections, is considered *P. hookeriana* Lehm. ssp. *chamissonis* Hult. The ssp. *hookeriana* is found in Northwest Greenland (4 coll. from 76°30′–77°30′N), North Greenland (2 coll. from 81°40′–82°35′N), and Northeast Greenland (3 coll. from 74°05–25′N). The taxon is mapped as *Potentilla hookeriana* Lehm. *s.l.*

Puccinellia bruggemanni Th. Sør.
The species described by Sørensen in Porsild (1955) is for the first time recognized in Greenland. It is distinguished from *Puccinellia angustata* (R. Br.) Rand. & Redf. by the glabrous branches of the panicle, the ovate spikelets and the incurved lemmas. Culms are usually only about 10 cm tall.

Ranunculus sabinei R. Br.
Two of the plants on a herbarium sheet with more collections from the Skærfjord-Store Koldewey area in Northeast Greenland collected and determined by Th. Sørensen as *Ranunculus sulphureus* are *R. sabinei.* The localities are head of Klægbugt (77°34′N), north side of C. F. Mourier Fjord (77°25′N), and east side of Store Koldewey north of Trækpasset (76°17′N). It is not clear, from the arrangement of the plants on the sheet and the labels, to which locality the plants belong. Consequently, the dot on the map which in any case marks the southernmost occurrence in Greenland cannot confidently be placed on the map number 83.

Salix arctica Pall./*S. glauca* L.
As hybridization between *Salix arctica* and *S. glauca* is possible (Skovortsov 1971) only fertile specimens have been included in the maps. The characteristics of the bracts (colour, hairiness) are used as the main diagnostic character.

Saxifraga caespitosa L. *s.l.*
This species includes three subspecies: 1) ssp. *caespitosa,* 2) ssp. *uniflora* (R. Br.) A. E. Pors., and 3) ssp. *laxiuscula* (Engl. & Irmsch.) Löve & Löve. Typical ssp. *caespitosa* are found in the area south of 78°10′N in West Greenland and 76°30′N in East Greenland. Ssp. *uniflora* are restricted to the high arctic Greenland. No

specimens have been determined to ssp. *laxiuscula*. The taxon is treated as one taxonomic unit when mapped.

Saxifraga hirculus L. *s.l.*
The arctic part of the species is usually distinguished as var. *propingua*. Böcher (1975) refers it to the high arctic distribution type but judging from the total distribution of the species (Hultén 1971) it should be considered as arctic widespread although the Greenland material is within the high arctic part of Greenland.

Saxifraga nivalis L./*S. tenuis* (Wbg.) H. Smith
Several authors consider *S. tenuis* a subspecies or variety of *S. nivalis* (Hultén 1968, 1971, Scoggan 1978). In Böcher *et al.* (1978) *S. tenuis* is treated on the species level, which is followed here. The hairiness of the stem is used as the main diagnostic character. Plants with dense, white curly hairs are *S. nivalis* and with short, purple hairs *S. tenuis*.

Stellaria longipes Goldie *s.l.*
The Greenland material of the taxon has been treated by Böcher (1951) and Philipp (1972). The *Stellaria longipes* complex is split up into six species of lower taxonomical value by some authors (Böcher *et al.* 1966): *S. longipes Goldie s.str., S. edwardsii* R. Br., *S. crassipes* Hult., *S. monantha* Hult., *S. laxmannii* Fish., and *S. laeta* Richards. Typical specimens of all species are recognized in the material from northern Greenland. However, all kind of intermediate forms, especially concerning the hairiness of the sepals and the type of bracts occur, and the material is treated as one taxon: *Stellaria longipes* Goldie *s.l.*

The distribution of typical specimens of the lower taxa is given below:

S. longipes: Mostly occurring in the southern part. *S. edwardsii:* Occurring in the whole area – rare in North Greenland. *S. laeta:* Not found in East Greenland south of Dove Bugt and only common in Thule distr. *S. laxmannii:* Only in district 1, 2 and 6. It is only common in

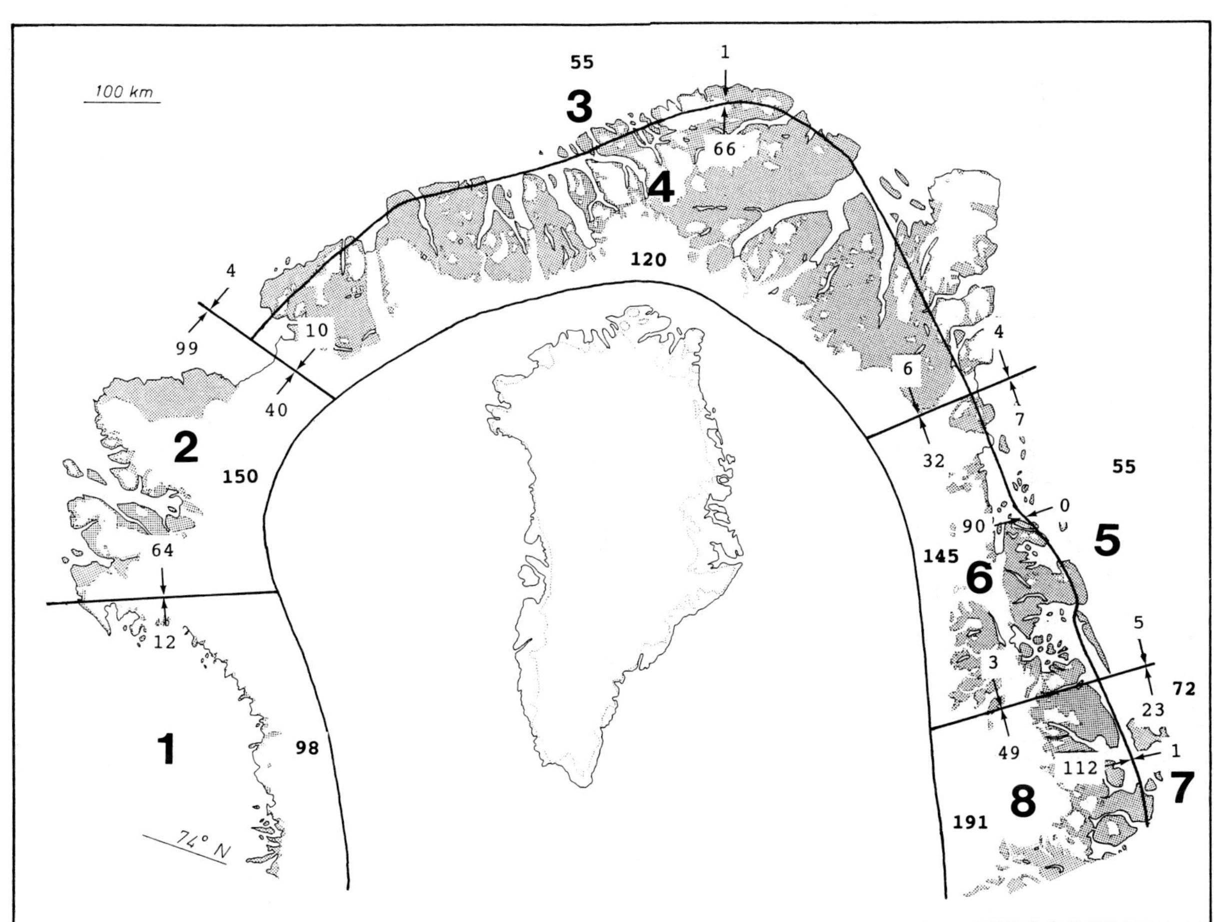

Fig. 8. Phytogeographical division of northern Greenland and number of taxa (small numbers in bold) in the districts. The number of distribution limits between the districts is indicated. E.g. the border between district 1 and 2 is placed where 12 southern species have their northern limit and 64 northern species their southern limit.

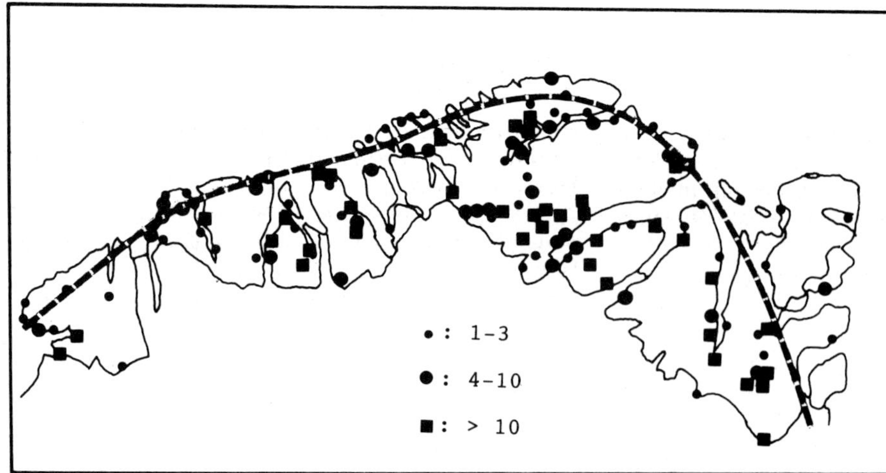

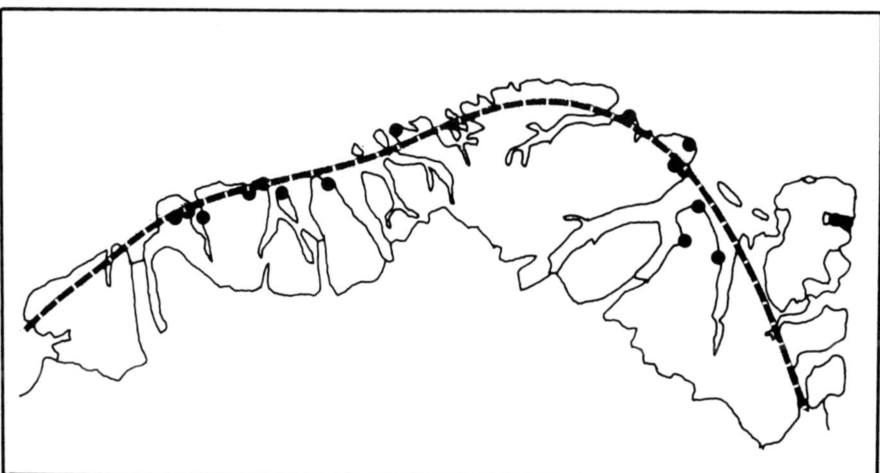

Fig. 9. Delimitation of the districts in the floristic province North Greenland. The distribution of 66 species which are almost restricted to the continental inland area (upper) and the only species *(Ranunculus sabinei)* exclusively occurring in coastal North Greenland (lower) are mapped. The number of inland species per locality is indicated.

district 2. *S. monantha:* Only in district 1, 2, and 3 (few). *S. crassipes:* The most common species throughout the area except for district 6. Especially common in North Greenland.

The variation of the complex is illustrated by a collection from Thule. The collection was considered one giant plant growing in a sand dune. Concerning the size of the specimens and the shape and size of the leaves, the colour of the plants and the onthogenetic age of the flowers, all the individuals in the collection are identical. However, there is a great variation in the hairiness of the sepals (glabrous-marginal hairy-hairy on the underside). The flowers are subtended by scale-like bracts with a membranous margin, or foliaceous at the apex, or by foliage leaves.

4.4. Distribution maps

For each taxon a distribution map is made showing the detailed distribution in the study area and the total Greenland distribution. The distribution maps only include specimens seen, as information from the literature about old collections is considered unreliable (Simmons 1904, 1909).

The Greenland distribution of the species based on information from Böcher *et al.* 1978, Feilberg 1984, Fredskild in prep., and Halliday in prep. in addition to herbarium material in C are given in appendix I.

4.5. Subdivision of the floristic province North Greenland

Based on coincidence in distribution of a large number of species exclusively or almost exclusively occurring in the interior North Greenland it has been possible to subdivide this province. These 66 species have been mapped together in addition to the only species (*Ranunculus sabinei*) which definitely prefers the outer coast areas (Fig. 9). From this map it is possible to draw a line indicating the demarcation between the two districts in North Greenland: 1) the interior North Greenland (N_i), and 2) the coastal North Greenland (N_c).

5. Results and discussion

5.1. Flora

Since Ostenfeld in 1925 optimistically stated that: "Flowering plants and ferns from North Greenland must be considered rather well known and further information will probably not add many species to the list" botanical work in the region has contributed to a more complete and detailed picture of the flora of North Greenland. Ostenfeld (1925a) estimated that 125 species of vascular plants would occur in North Greenland – north of 76°N, including the districts: Thule, North Greenland, and Northeast Greenland. Today a total of 172 taxa is identified from this area; an increase in number of 38%. 121 taxa are known from the floristic province North Greenland north of 79°30′N as defined by Böcher *et al.* (1959). The total number of taxa from Northwest and Northeast Greenland between 74°N and 79°30′N is 161 and 194, respectively.

The variation in the species diversity in the huge study area is very large. Thus well-investigated sites in the comparatively luxuriant inland district of North Greenland have c. 80 species against only 40 on the most harsh outer coast localities. The number of species in inland and coastal localities in Northeast Greenland is app. 110 and 50–70, respectively. The highest number of species is 144 at Zackenberg in Wollaston Forland. Fig. 8 shows the number of species in the districts. Generally, species diversity declines when moving to the north except for Melville Bugt. The lower number of species in Melville Bugt compared to Thule district must be ascribed both to climatic and edaphic conditions. Some species are rare in western North Greenland or are completely missing in the area from Hall Land to Wulff Land and this reduction in the species diversity can only partly be explained by the fact that this part of North Greenland has been less intensively investigated.

5.2. Phytogeography

5.2.1. Phytogeographical districts

Based on phytogeographical borders which appear when comparing the distribution types, northern Greenland has been divided into eight districts (Fig. 8). Important distribution limits occur at 76°N and 79°30′N in both West and East Greenland. The number of taxa to which the border marks a distribution limit is given on each side of the border line. The border lines at 76°N and 79°30′N are based on a large number of limits of distribution – between 38 and 76 with the highest values in West Greenland making Thule district the most outstanding phytogeographical district. 67 species have a limit of distribution between the two districts in North Greenland.

The delimitation of the two North Greenland districts is based on the distribution of taxa with 1) a continuous distribution in the whole of or part of West and/or East Greenland in addition to occurrences in the interior parts of North Greenland and 2) the species which definitely prefer the outer coast area. Mapped together it is possible to draw a line between the coastal and the interior North Greenland districts (Fig. 9). Because of fewer well-investigated localities in western North Greenland and the impoverishment of the flora in this area the delimitation – especially in Washington Land – is not as well-founded as in the eastern part. The 67 species used for the delimitation include species from the types: 1b, 1c, 2b, 4a, 4b, 5a, 6a, 6c, 7a, and 8a. Of 835 collections only 41 (4,9%) are from outside the district in which they have their main distribution. The only distinct outer coast species is *Ranunculus sabinei,* which was only found few times in an inland locality close to the border line.

The eight districts are:

1. Melville Bugt district: from 74°N to Crimson Cliffs (app. 76°N),
2. Thule district: Crimson Cliffs – Humboldt Gletscher (app. 79°N),
3. coastal North Greenland,
4. interior North Greenland,
5. coastal parts of northern Northeast Greenland: Bessel Fjord (76°N) – Nioghalvfjerdsfjorden (app. 79°30′N),
6. interior parts of northern Northeast Greenland,
7. coastal parts of southern Northeast Greenland: Gael Hamkes Bugt (app. 74°N) – Bessel Fjord, and
8. interior parts of southern Northeast Greenland.

5.2.2. Distribution types

The distribution maps of the 218 taxa have been classified into 14 North Greenland distribution types (NGDT) and 16 subtypes (Table 4 and 5). The presence or absence of the taxa in the district is used as the primary basis for classifying the distribution maps. In addition, their frequency in the districts has been used. As long as the main part of the specimens are from localities within the districts defining the type, it is accepted that one, two or three collections occur in the neighbouring districts. Table 5 gives the definition of the distribution types, illustrated in Fig. 10.

Distribution type 1 includes species which occur throughout the region. Type 2 and 3 have a gradual reduction in the distribution in western North Greenland, whereas type 4 is high arctic occurring only north of 76°N. Type 5–8 include southern taxa with a marked distribution limit at 79°30′N either at both coasts (type 5 and 8) or only at one coast (type 6 and 7). Type 9 has a distribution limit at 79°30′N and 76°N in West and East Greenland, respectively. Taxa restricted to the area between 76° and 79°30′N on the west and east coast are referred to type 10 and 11, respectively. Type 12–14 include types of low arctic taxa with a marked distribu-

Table 4. Definition of the North Greenland distribution types (NGDT).

1. In all districts.
1a. Like 1 but missing in southern parts of coastal Northeast Greenland.
1b. Like 1 but missing in coastal North Greenland.
1c. Like 1b but in addition missing in parts of coastal Northeast Greenland.
2. In all districts except Melville Bugt.
2a. Like 2 but missing in parts of coastal Northeast Greenland.
2b. Like 2a and in addition missing in coastal North Greenland.
3. In North and Northeast Greenland.
3a. Like 3 but missing in parts of coastal Northeast Greenland.
4. North of 76°N in both West and East Greenland.
4a. Like 4 but in North Greenland only in the coastal part.
4b. Like 4 but in North Greenland only in the interior part and in addition missing in district 5 and 6.
5. South of 79°30′N in both West and East Greenland but missing in parts of coastal Northeast Greenland.
5a. Like 5 and in addition isolated occurrences in the interior North Greenland.
6. South of 79°30′N in West Greenland.
6a. Like 6 and in addition isolated occurrences in the interior North Greenland.
6b. Like 6 and in addition south of 76°N in East Greenland.
6c. Like 6b plus isolated occurrences in the interior North Greenland.
7. South of 79°30′N in East Greenland but missing in parts of coastal Northeast Greenland.
7a. Like 7 plus isolated occurrences in the interior North Greenland.
8. South of 79°30′N but missing in Melville Bugt.
8a. Like 8 plus isolated occurrences in the interior North Greenland and missing in parts of coastal Northeast Greenland.
9. South of 79°30′N in West Greenland and 76°N in East Greenland but missing in Melville Bugt.
10. Only in Northwest Greenland (district 2).
11. Only in Northeast Greenland (district 6).
12. South of 76°N in both West and East Greenland.
13. South of 76°N in West Greenland.
14. South of 76°N in East Greenland but only in the interior parts.
14a. South of 76°N in East Greenland.

tion limit at 76°N either at both coasts (type 12) or only at one (type 13 and 14).

In order to relate the local distribution in northern Greenland with the total distribution of the species NGDT are compared with the Greenland and holarctic distribution (Table 6) and the biological distribution types according to Böcher (1975) (Table 7), which is a revised, Danish version of the classification proposed in Böcher *et al.* (1959). The conclusions from these tables are discussed in connection with the description of the types given below.

The list of taxa in northern Greenland is given in Table 8 with statement of their biological and geographical distribution and numbers referring to the dot maps.

5.2.3. Definition and description of the distribution types

Table 4 and 5 outline the definition and number of taxa in the distribution types. A synoptical diagram of the distribution types is given in Fig. 10.

Type 1 (Map Nos 1–32)
This type includes taxa that are evenly distributed throughout the study area. Some of the taxa are seemingly rare in western North Greenland, which may be explained by the lesser activity in this area. Most likely

Table 5. Distribution of taxa in the districts and number of taxa within the distribution types. A hyphen between the numbers of the districts indicates a continuous distribution in these areas. Map numbers are given.

Distribution type	Districts	No. of taxa	Map No.
1	1–8	32	1–32
1a	1–6, 8	1	33
1b	1–2, 4–8	5	34–38
1c	1–2, 4, 5–6, 8	4	39–42
	1–2, 4, 6–8	2	43–44
	1–2, 4, 6, 8	6	45–50
2	2–8	8	51–58
2a	2–6, 8	1	59
	2–4, 6–8	3	60–62
	2–4, 6, 8	4	63–66
2b	2, 4, 6–8	5	67–71
	2, 4, 6, 8	6	72–77
3	3–8	1	78
3a	3–4, 6–8	2	79–80
	3–4, 6, 7	1	81
4	2–4, 6	1	82
4a	2–3, 6	1	83
4b	2, 4	4	84–87
5	1–2, 6–8	1	88
	1–2, 6, 8	14	89–102
5a	1–2, 4, 5–8	1	103
	1–2, 4, 6–8	1	104
	1–2, 4, 6, 8	7	105–111
6	1–2	5	112–116
6a	1–2, 4	1	117
6b	1–2, 8	5	118–122
6c	1–2, 4, 8	1	123
7	6–8	2	124–125
	6, 8	7	126–132
7a	4, 6–8	2	133–134
	4, 6, 8	4	135–138
7b	1, 6, 8	1	139
8	2, 6–8	1	140
	2, 6, 8	3	141–143
8a	2, 4, 5–8	2	144–145
	2, 4, 6–8	2	146–147
	2, 4, 6, 8	13	148–160
9	2, 8	1	161
10	2	9	162–170
11	6	1	171
12	1, 8	6	172–177
13	1	5	178–182
14	8	35	183–217
14a	7, 8	1	218

Table 6. The percentage composition of the geographical distribution types according to Hultén (1958, 1962, 1971) in the North Greenland distribution types (NGDT). Cir: circumpolar, Amp-A: amphi-Atlantic, W: western distribution i.e. main occurrence in North America, E: eastern distribution i.e. main occurrence in Eurasia, Amp-B: amphi-Beringian, G: endemic to Greenland.

NGDT	Cir	Amp-A	W	E	Amp-B	G	n
1	76	8	16				50
2	55	15	26	4			27
3	25		50	25			4
4	17		83				6
5	75	13	13				24
6	58	8	33				12
7	25	19	31	25			16
8	76	5	19				21
9		100					1
10	44		56				9
11					100		1
12	33	17	33	17			6
13		80	20				5
14	44	28	6	19		3	36
n	122	32	48	14	1	1	218
(%)	56	15	22	6	0,5	0,5	

Cardamine bellidifolia (Map 1), *Saxifraga hyperborea* (Map 16), and other species are growing in the area.

The Greenland distribution of the taxa falls into two distinct groups. 66% have a circumgreenlandic distribution (*Draba lactea*, Map 21, differs in being missing in Southeast Greenland) and 34% have a high arctic distribution with no occurrence in southern Greenland. Most of the species (76%) have a wide circumpolar distribution.

Half of the species are classified as widespread arctic species according to Böcher (1975) while the other half is mainly arctic-continental or high arctic. Only four species are known to have their main distribution in low arctic and boreal areas.

Subtype 1a (Map No. 33)

Dryas integrifolia is the only species in this subtype differing from the main type by being missing in coastal Northeast Greenland.

It is a western taxon classified as arctic-continental according to Böcher (1975).

Subtype 1b (Map Nos 34–38)

This subtype differs from the main type in being restricted, in North Greenland, to the interior parts.

Four of the five species have a circumgreenlandic distribution. *Melandrium triflorum* (Map 38) differs by being missing in southern Greenland.

Subtype 1c (Map Nos 39–50)

This subtype includes species which, in addition to subtype 1b, are missing from one or both of the coastal districts in Northeast Greenland.

Eight species are known from all parts of Greenland. *Melandrium affine* (Map 41), *Lesquerella arctica* (Map 49), and *Potentilla hookeriana* (Map 50) are not known from South and Southeast Greenland. *Carex stans* (Map 42) has the most restricted distribution; not known south of 74°N in Northeast Greenland.

Type 2 (Map Nos 51–58)

This type includes species that are widespread in the study area but missing in Melville Bugt.

Concerning the Greenland distribution of the main type and the two subtypes the species fall into three groups: 1) distributed throughout Greenland except for

Table 7. The percentage composition of the biological distribution types according to Böcher (1975) in the North Greenland distribution types. The abbreviations, see Table 8.

NGDT	A	HA	AC	MA	L	LO	LC	B	BC	n
1	52	16	26	2			2	2		50
2	7	78	11		4					27
3		100								4
4		83	17							4
5	21	4	17	21	25	8	4			24
6			50	8	25			17		12
7	6	13	19	19	19	13	6	6		16
8	10	10	19	24	14		14	10		21
9				100						1
10		22	11		11		33	11	11	9
11	100									1
12					50	33	17			6
13					60	40				5
14			6	19	17	31	17	11		36
n	37	45	37	23	29	19	16	11	1	218
(%)	17	21	17	11		29		5		

Table 8. List of taxa present in northern Greenland. The numbers of the distribution maps and the distribution types are given. Three types of codes concerning the distribution patterns are given. 1) North Greenland distribution type (NGDT), 2) climatic distribution type (CDT) according to Böcher (1975), 3) Holarctic distribution type (HDT) according to Hultén (1958, 1962, 1971), Hultén & Fries (1986), Porsild (1964), and Feilberg (1984).

	Map no.	NGDT	CDT	HDT
Alopecurus alpinus Sm.	22	1	HA	Cir
Androsace septentrionalis L.	167	10	AC	Cir
Antennaria canescens (Lge.) Malte	172	12	LO	W
Antennaria compacta Malte (see *A. ekmaniana*)				
Antennaria ekmaniana A. E. Pors.	114	6	AC	W
Antennaria porsildii Ekm.	203	14	MA	Amp-A
Arabis alpina L.	183	14	LO	Amp-A
Arabis arenicola (Richards.) Gel.	201	14	MA	W
Arctagrostis latifolia (R. Br.) Griseb.	60	2a	HA	Cir
Arctostaphylos alpina (L.) Spreng.	204	14	L	Cir
Arenaria humifusa Wbg.	161	9	MA	Amp-A
Arenaria pseudofrigida (Ostf. & Dahl) Juz.	133	7a	HA	E
Armeria scabra Pall. ssp. *sibirica* (Turcz.) Hyl.	146	8a	AC	Cir
Arnica angustifolia M. Vahl in Horn.	109	5a	AC	Amp-A
Betula nana L.	173	12	LC	E
Braya humilis (C. A. Mey.) Robins.	137	7a	AC	W
Braya intermedia Th. Sør. (see *B. humilis*)				
Braya linearis Rouy	205	14	LC	Amp-A
Braya purpurascens (R. Br.) Bge.	63	2a	HA	Cir
Braya thorild-wulffii Ostf.	82	4	HA	W
Calamagrostis neglecta (Ehrh.) Gaertn., Mey & Scherb.	184	14	B	Cir
Calamagrostis purpurascens R. Br.	155	8a	AC	W
Campanula gieseckiana Vest *s.l.*	174	12	L	Cir
Campanula uniflora L.	104	5a	MA	Amp-A
Cardamine bellidifolia L.	1	1	A	Cir
Cardamine pratensis L. *s.l.*	148	8a	L	Cir
Carex atrofusca Schkuhr	156	8a	MA	Cir
Carex bigelowii Torr. *s.l.*	89	5	L	Cir
Carex boecheriana Löve, Löve & Raymond	131	7	LC	W
Carex capillaris L. ssp. *fuscidula* (Krecz.) Löve & Löve	105	5a	L	Cir
Carex glacialis Mack.	123	6c	AC	Cir
Carex glareosa Wbg.	118	6b	L	Cir
Carex lachenalii Schkuhr	119	6b	L	Cir
Carex marina Dew. ssp. *pseudolagopina* (Th. Sør.) Böch.	157	8a	LC	Cir
Carex maritima Gunn.	67	2b	A	Cir
Carex microglochin Wbg.	185	14	LC	Amp-A
Carex misandra R. Br.	23	1	AC	Cir
Carex nardina Fr.	2	1	AC	Amp-A
Carex norvegica Retz. *s.l.*	186	14	LO	E
Carex parallela (Læst.) Sommerf.	208	14	LC	E
Carex rariflora (Wbg.) Sm.	187	14	L	Cir
Carex rupestris All.	145	8a	AC	Cir
Carex saxatilis L.	149	8a	A	Cir
Carex scirpoidea Michx.	175	12	L	W
Carex stans Drej.	42	1c	HA	Cir
Carex subspathacea Wormsk.	126	7	L	Cir
Carex supina Wbg. ssp. *spaniocarpa* (Steud.) Hult.	90	5	LC	Cir
Carex ursina Dew.	111	5a	MA	Cir
Carex vaginata Tausch.	209	14	B	Cir
Cassiope tetragona (L.) D. Don.	24	1	AC	Cir
Cerastium alpinum L. ssp. *lanatum* (Lam.) Asch. & Gaerbn. (see *C. arcticum*)	24	1	AC	Cir
Cerastium arcticum Lge. *s.l.*	3	1	A	Amp-A
Cerastium cerastoides (L.) Britt.	188	14	LO	Amp-A
Cerastium regelii Ostf. ssp. *caespitosum* (Malmgr.) Tolm.	78	3	HA	Cir
Chamaenerion latifolium (L.) Sweet	45	1c	A	W
Chrysosplenium tetrandrum (N. Lund) Th. Fr.	210	14	MA	Cir
Cochlearia groenlandica L.	4	1	A	Cir
Colpodium vahlianum (Liebm.) Nevski	51	2	HA	Cir
Cystopteris fragilis (L.) Bernh. *s.l.*	46	1c	AC	Cir
Deschampsia brevifolia R. Br.	70	2b	HA	Cir
Deschampsia pumila Ostf.	101	5	HA	Cir
Diapensia lapponica L. ssp. *lapponica*	112	6	L	Amp-A

Continue

	Map no.	NGDT	CDT	HDT
Draba adamsii Led.	52	2	HA	Cir
Draba alpina L.	144	8a	MA	Cir
Draba arctica J. Vahl.	68	2b	AC	(Cir)
Draba arctogena Ekm.	65	2a	HA	W
Draba bellii Holm	53	2	HA	W
Draba cinerea Adams (see *D. arctica*)				
Draba crassifolia Grah.	200	14	MA	Amp-A
Draba fladnizensis Wulf.	139	7b	A	E
Draba gredinii Ebm. (see *D. alpina*)				
Draba glabella Pursh	102	5	AC	Cir
Draba lactea Adams	21	1	A	Cir
Draba macrocarpa Adams (see *D. bellii*)				
Draba micropetala Hook. (see *D. adamsii*)				
Draba nivalis Liljebl.	91	5	A	Cir
Draba norvegica Gunn. (see *D. glabella)*				
Draba oblongata R.Br. (see *D. adamsii*)				
Draba subcapitata Simm.	54	2	HA	Cir
Dryas integrifolia M. Vahl	33	1a	AC	W
Dryas octopetala L. ssp. *punctata* (Juz.) Hult.	124	7	HA	E
Dryopteris fragrans (L.) Schott	115	6	AC	Cir
Dupontia fisheri R. Br.	168	10	HA	Cir
Dupontia psilosantha Rupr.	206	14	MA	Cir
Eleocharis acicularis (L.) R. & S.	162	10	B	Cir
Elymus hyperarcticus (Polun.) Tzvel.	135	7a	AC	W
Empetrum nigrum L. ssp. *hermaphroditum* (Hagerup) Böch.	92	5	L	Cir
Epilobium arcticum Sam.	136	7a	AC	Amp-A
Equisetum arvense L. *s.l.*	47	1c	B	Cir
Equisetum variegatum Schleich.	74	2b	L	Cir
Erigeron compositus Pursh	72	2b	AC	W
Erigeron eriocephalus J. Vahl.	154	8a	MA	Cir
Erigeron humilis Grah.	130	7	MA	W
Eriophorum angustifolium Honck. ssp. *subarcticum* (V. Vassil.) Hult.	113	6	B	Cir
Eriophorum callitrix Cham.	160	8a	HA	Cir
Eriophorum scheuchzeri Hoppe	34	1b	A	Cir
Eriophorum triste (Th. Fr.) Hadac & Löve	55	2	HA	Cir
Erysimum pallasii (Pursh) Fern.	84	4b	HA	Cir
Euphrasia frigida Pugsl.	127	7	L	Amp-A
Eutrema edwardsii R. Br.	76	2b	HA	Cir
Festuca baffinensis Polun.	75	2b	HA	W
Festuca brachyphylla Schult. & Schult.	39	1c	A	Cir
Festuca hyperborea Holmen ex Frederiksen	25	1	HA	W
Festuca rubra L. *s.l.*	189	14	L	Cir
Festuca vivipara (L.) Sm. ssp. *glabra* Frederiksen	134	7a	LO	Cir
Gentiana tenella Rottb.	132	7	L	E
Geum rossii (R. Br.) Ser.	171	11	A	Amp-B
Halimolobus mollis (Hook.) Rollins	165	10	LC	W
Harrimanella hypnoides (L.) Coville	176	12	LO	Amp-A
Hierochloë alpina (Willd.) R. & S.	5	1	AC	Cir
Hippuris vulgaris L. *s.l.*	150	8a	B	Cir
Honckenya peploides (L.) Ehrh. var. *diffusa* (Horn.) Mattf.	141	8	B	Cir
Huperzia selago (L.) Bernh. ex Schrank & Mart. ssp. *arctica* (Grossh.) Löve & Löve	106	5a	A	Cir
Juncus arcticus Willd.	190	14	L	Cir
Juncus biglumis L.	6	1	A	Cir
Juncus castaneus Sm.	110	5a	MA	Cir
Juncus trifidus L.	191	14	LO	Amp-A
Juncus triglumis L. *s.l.*	73	2b	A	Cir
Kobresia myosuroides (Vill.) Fiori & Paol.	43	1c	LC	Cir
Kobresia simpliciuscula (Wbg.) Mack.	153	8a	LC	Cir
Koenigia islandica L.	107	5a	A	Cir
Ledum palustre L. ssp. *decumbens* (Ait.) Hult.	166	10	LC	W
Lesquerella arctica (Wormsk.) S. Wats.	49	1c	AC	W
Loiseleuria procumbens (L.) Desv.	178	13	LO	Amp-A
Luzula arctica Blytt	26	1	AC	Cir
Luzula confusa Lindeb.	7	1	LO	Cir
Luzula spicata (L.) DC	128	7	LO	Cir

Continue

	Map no.	NGDT	CDT	HDT
Luzula wahlenbergii Rupr.	211	14	L	Cir
Matricaria maritima L. ssp. *phaeocephala* (Rupr.) Rauschert	212	14	AC	Cir
Melandrium affine J. Vahl *s.l.*	41	1c	MA	W
Melandrium apetalum (L.) Fenzl ssp. *arcticum* (Fr.) Hult.	56	2	HA	Cir
Melandrium triflorum (R. Br.) J. Vahl	38	1b	AC	W
Mertensia maritima (L.) S. F. Gray	163	10	L	W
Minuartia biflora (L.) Sch & Th.	93	5	LO*	Cir
Minuartia rossii (R. Br.) Graebn.	66	2a	HA	W
Minuartia rubella (Wbg.) Hiern	8	1	A	Cir
Minuartia stricta (Sw.) Hiern	142	8	MA	Cir
Oxyria digyna (L.) Hill	9	1	A	Cir
Papaver radicatum Rottb. *s.l.*	10	1	A	Cir
Pedicularis capitata Adams	85	4b	AC	W
Pedicularis flammea L.	94	5	L	W
Pedicularis hirsuta L.	27	1	AC	Amp-A
Pedicularis langsdorfii Fisch. ex Stev. ssp. *arctica* (R. Br.) Pennell	169	10	AC	W
Pedicularis lapponica L.	202	14	LC	Cir
Pedicularis sudetica Willd. ssp. *albolabiata* Hult.	170	10	HA*	W
Phippsia algida (Sol.) R. Br.	103	5a	A	Cir
Phippsia algida (Sol.) R. Br. ssp. *algidiformis* (H. Sm.) L. & L.	59	2a	HA*	E
Phyllodoce coerulea (L.) Bab.	179	13	LO	Amp-A
Pleuropogon sabinei R. Br.	71	2b	HA	Amp-A
Poa abbreviata R. Br.	57	2	HA	Amp-A
Poa alpina L.	192	14	LO	Amp-A
Poa alpina L. var. *vivipara* L.	213	14	(LO)	E
Poa arctica R. Br.	11	1	AC	Cir
Poa glauca M. Vahl	35	1b	A	Cir
Poa hartzii Gandoger *s.l.*	77	2b	AC	W
Poa pratensis L. *s.l.*	120	6b	B	Cir
Poa pratensis L. var. *colpodea* (Th. Fr.) Schol.	64	2a	HA	Cir
Polemonium boreale Adams	218	14	AC	Cir
Polygonum viviparum L.	12	1	A	Cir
Potamogeton filiformis Pers.	129	7	B	Amp-A
Potentilla crantzii (Cr.) G. Beck	193	14	LO	Amp-A
Potentilla hookeriana Lehm. *s.l.*	50	1c	AC	Cir
Potentilla hyparctica Malte	28	1	HA	Cir
Potentilla nivea L. emend. Hult.	95	5	AC	Cir
Potentilla pulchella R. Br.	61	2a	HA	Amp-A
Potentilla rubella Th. Sør.	214	14	(MA)	E
Potentilla rubricaulis Lehm.	69	2b	HA	W
Potentilla stipularis L.	215	14	LC	E
Potentilla vahliana Lehm.	116	6	AC	W
Primula stricta Horn.	207	14	LC	W
Puccinellia andersonii Swall.	158	8a	HA	W
Puccinellia angustata (R. Br.) Rand. & Redf.	58	2	HA	Cir
Puccinellia bruggemanni Th. Sør.	79	3a	HA*	W
Puccinellia coarctata Fern. & Weath.	194	14	L	Amp-A
Puccinellia phryganodes (Trin.) Schribn. & Merr.	48	1c	A	Cir
Puccinellia vaginata (Lge.) Fern. & Weath.	122	6b	MA	W
Pyrola grandiflora Rad.	121	6b	AC	Cir
Ranunculus affinis R. Br. *s.l.*	159	8a	LC	W
Ranunculus confervoides (Fr.) Fr.	152	8a	L	Cir
Ranunculus glacialis L.	125	7	MA	E
Ranunculus hyperboreus Rottb.	36	1b	A	Cir
Ranunculus nivalis L.	147	8a	MA	Cir
Ranunculus pygmaeus Wbg.	88	5	MA	Cir
Ranunculus sabinei R. Br.	83	4a	HA	W
Ranunculus sulphureus Sol. in Phipps	29	1	HA	Cir
Rhodiola rosea L.	195	14	LO	E
Rhododendron lapponicum (L.) Wbg.	96	5	AC	W
Rumex acetosella L. *s.l.*	196	14	B	Cir
Sagina caespitosa (J. Vahl) Lge.	100	5	MA	W
Sagina intermedia Fenzl	13	1	A	Cir
Salix arctica Pall.	30	1	HA	Cir
Salix glauca L. *s.l.*	180	13	L	W
Salix herbacea L.	97	5	LO	Amp-A
Saxifraga aizoides L.	151	8a	L	Amp-A

Continue

	Map no.	NGDT	CDT	HDT
Saxifraga caespitosa L. *s.l.*	14	1	A	Cir
Saxifraga cernua L.	15	1	A	Cir
Saxifraga foliolosa R. Br.	31	1	HA	Cir
Saxifraga hieracifolia W. & K.	138	7a	MA	Cir
Saxifraga hirculus L. *s.l.*	140	8	A*	Cir
Saxifraga hyperborea R. Br.	16	1	HA	Cir
Saxifraga nathorstii (Dusén) Hayek	216	14	MA	G
Saxifraga nivalis L.	17	1	A	Cir
Saxifraga oppositifolia L.	18	1	A	Cir
Saxifraga paniculata Mill.	181	13	L	Amp-A
Saxifraga platysepala (Trautv.) Tolm.	62	2a	HA	Amp-A
Saxifraga rivularis L.	98	5	L	Cir
Saxifraga tenuis (Wbg.) H. Smith	19	1	A	Cir
Saxifraga tricuspidata Rottb.	117	6a	AC	W
Sibbaldia procumbens L.	197	14	LO	Cir
Silene acaulis (L.) Jacq.	37	1b	A	Amp-A
Stellaria humifusa Rottb.	99	5	A	Cir
Stellaria longipes Goldie *s.l.*	20	1	(AC)	(W)
Taraxacum arcticum (Trautv.) Dahlst.	80	3a	HA	E
Taraxacum arctogenum Dahlst.	86	4b	HA	W
Taraxacum brachyceras Dahlst.	217	14	(LO)	E
Taraxacum hyparcticum Dahlst.	87	4b	HA	W
Taraxacum phymatocarpum J. Vahl	32	1	HA	W
Taraxacum pumilum Dahlst.	81	3a	HA	W
Thalictrum alpinum L.	198	14	LO	Cir
Tofieldia coccinea Richards.	143	8	AC	W
Tofieldia pusilla (Michx.) Pers.	177	12	L	Cir
Triglochin palustre L.	199	14	B	Cir
Trisetum spicatum (L.) Richt.	44	1c	A	Cir
Vaccinium uliginosum L.	108	5a	L	Cir
Vaccinium vitis-idaea L. ssp. *minus* (Lodd.) Hult.	164	10	BC	Cir
Woodsia alpina (Bolt.) S. F. Gray	182	13	L	Amp-A
Woodsia glabella R. Br.	40	1c	A	Cir

Abbreviations

Biological distribution types:

A: Arctic, widespread distribution
AC: Arctic, continental distribution
B: Boreal distribution
BC: Boreal, continental distribution
HA: High arctic distribution
L: Low arctic, widespread distribtion
LC: Low arctic, continental distribution
LO: Low arctic, continental distribution
MA: Middle arctic distribution

Holarctic distribution types:

Amp-A: Amphi-Atlantic distribution
Amp-B: Amphi-Beringian distribution
Cir: Circumpolar distribution
E: Eastern distribution, mainly in Eurasia
G: Endemic to Greenland
W: Western distribution, mainly in North America-Greenland

*: Distribution type proposed by the author
(): Uncertainty about the classification

Melville Bugt (14%), 2) high arctic distribution, with southern limit at 70°N both in West and East Greenland (64%), 3) high arctic distribution differing from the above-mentioned by not occurring south of 76°N in West Greenland (22%). Three of the species (*Saxifraga platysepala* (Map 62)), *Draba arctogena* (Map 65), and *Minuartia rossii* (Map 66) do not reach Scoresby Sund in East Greenland.

The main part (78%) belongs to the high arctic biological type. 55% are circumpolar and the rest occurs mostly (26%) in the western and amphi-Atlantic (15%) part of the Arctic. Eastern species constitute only 4%.

Subtype 2a (Map Nos 59–66)

This subtype differs from the main type by being missing in parts of coastal Northeast Greenland.

All the species have a high arctic distribution in Greenland not reaching further south than app. 69°N except for *Potentilla pulchella* (Map 61) which has been found as far south as 66°N.

Subtype 2b (Map Nos 67–77)

This subtype includes species, which in addition to the definition of type 2, are missing both in coastal parts of North and Northeast Greenland.

Their Greenland distribution falls into three groups: 1) circumgreenlandic (4 spp.), 2) high arctic (5 spp.), and 3) high arctic but only on the east coast (2 spp.).

Type 3 (Map No. 78)

This type and subtype 3a include taxa with a distribution only in North and Northeast Greenland, north of 79°30′N at the west coast. Two of the four species have southern limits in Scoresby Sund (70°N), while two only

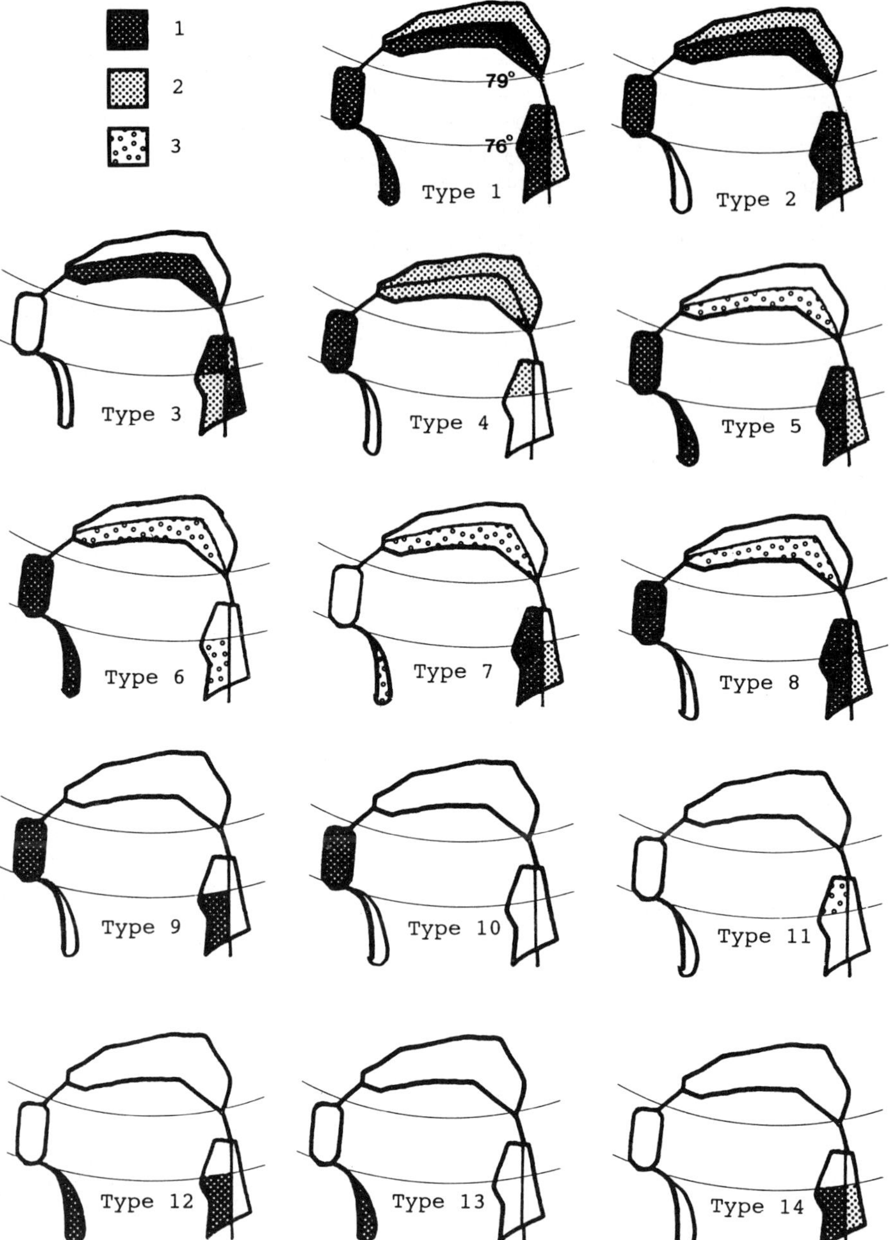

Fig. 10. Distribution types in northern Greenland. The hatched and dotted areas indicate that the main type includes subtypes. 1: continuous distribution, 2: a few species missing, 3: isolated occurrences of a few species.

reach 75°N in East Greenland. All the species are high arctic and half of them are mainly distributed in the western part of the Arctic.

Subtype 3a (Map Nos 79–81)
The three species in this subtype are, in addition to the definition of type 3, missing in one (*Puccinellia bruggemanni,* Map 79, and *Taraxacum arcticum,* Map 80), or two districts (*Taraxacum pumilum,* Map 81). *Taraxacum arcticum* reaches Scoresby Sund, while the others have a southern limit at 74–75°N.

Type 4 (Map No. 82)
With one exception all taxa of this type, including two subtypes, have their total Greenland distribution north of 76°N. *Braya thorild-wulffii* has isolated occurrences in the Disko-Nugssuaq area (70°N).

All the taxa belong either to the high arctic (83%) or the arctic-continental (17%) type, and it is the only type with no representatives of the low arctic types. 83% have their main distribution in North America and adjacent areas.

Subtype 4a (Map No. 83)
This subtype differs from the main type by occurring almost exclusively in the coastal part of North Greenland. *Ranunculus sabinei* is the only species in the type.

Subtype 4b (Map Nos 84–87)
This subtype includes species that are restricted to the interior parts of North Greenland. Three species are missing in western North Greenland. *Pedicularis capitata* (Map 85) has the most restricted distribution, only occurring in Thule district and central Peary Land.

Type 5 (Map Nos 88–102)
This type comprises taxa with a more or less continuous distribution northwards to 79°30′N in both West and East Greenland. Most of the species are restricted to the interior parts of Northeast Greenland.

Twelve of the 15 taxa have a low arctic distribution in Greenland from South Greenland to Humboldt Gletscher. *Deschampsia pumila* (Map 101) and *Draba glabella* (Map 102) differ in being missing from the southernmost and *Sagina caespitosa* (Map 100) from the southeastern part.

The type comprises a heterogenous group with taxa from seven of the biological distribution types. More than half of them are classified as low arctic or middle arctic. The main part (75%) has a circumpolar distribution.

Subtype 5a (Map Nos 103–111)
This subtype includes taxa that have additional occurrences in the interior North Greenland. Most are missing from one or two coastal districts in Northeast Greenland.

Six taxa out of the nine have a circumgreenlandic distribution except for parts of North Greenland. *Arnica angustifolia* (Map 109), *Juncus castaneus* (Map 110), and *Carex ursina* (Map 111) are not known from the southern parts of Greenland.

Type 6 (Map Nos 112–116)
This type consists of taxa only known from West Greenland south of 79°30′N.

Three of the five taxa are only known from West Greenland or part of it, while *Diapensia lapponica* (Map 112) reaches Jameson Land (71°N) in East Greenland. *Dryopteris fragrans* (Map 115) has a very limited distribution in East Greenland occurring only in the interior parts of Scoresby Sund.

Half of the taxa are arctic-continental and 25% are low arctic. This main type has the largest number of boreal species (17%). Unexpectedly only 33% are western species; the main part has a circumpolar distribution.

Subtype 6a (Map No. 117)
This subtype includes one species which is known from the west coast with an isolated occurrence in the interior North Greenland.

Saxifraga tricuspidata (Map 117) has been found a few times in the interior parts of Peary Land and Washington Land. Its southern limit is at app. 64°N in West Greenland. In addition it has isolated occurrences at the entrance of Scoresby Sund.

Subtype 6b (Map Nos 118–122)
This subtype consists of species that occur on the west coast to app. 79°N and to app. 76°N at the east coast. Four of the five species are low arctic with a continuous distribution in South Greenland. Only *Puccinellia vaginata* (Map 122) differs in not occurring south of app. 67°N.

Subtype 6c (Map No. 123)
This subtype is a combination of 6a and 6b, and includes only *Carex glacialis,* which is distributed northwards to 78°N in West Greenland and to 76°N in East Greenland, with isolated occurrences in the interior of North Greenland.

Type 7 (Map Nos 124–132)
This type includes taxa that are distributed northwards to 79°N in East Greenland but missing in one (*Dryas octopetala* (Map 124)) or two coastal districts in Northeast Greenland.

Five of the nine species are widespread in low arctic Greenland and do not reach as far north in West as in East Greenland. The northern limit in West Greenland is at app. 72°N. *Gentiana tenella* (Map 132) and *Carex boecheriana* (Map 131) occur in a limited area on both coasts, while *Ranunculus glacialis* (Map 125) and *Dryas octopetala* are only known from East Greenland.

This type includes the widest spectrum of biological distribution types with species of eight of the nine types. Low arctic types comprise only 38%. Arctic-continental and middle arctic species have an equal frequency of 19%. One third is distributed in western Arctic and one fourth is circumpolar or distributed mainly in Eurasia.

Subtype 7a (Map Nos 133–138)
This subtype includes taxa which have in addition isolated occurrences in the interior North Greenland. *Arenaria pseudofrigida* (Map 133), *Braya humilis* (Map 137), *Festuca vivipara* (Map 134), and *Saxifraga hieracifolia* (Map 138) have their main distribution in East Greenland, while *Epilobium arcticum* (Map 136) and *Elymus hyperarcticus* (Map 135) in addition are known from the Disko-Nuussuaq area in West Greenland.

Subtype 7b (Map No. 139)
Draba fladnizensis is the only species in this subtype defined by having northern limit at 76°N in West and at 79°N in East Greenland. It has southern limits at 66°N in West and at 70°N in East Greenland.

Type 8 (Map Nos 140–143)
This type includes taxa which reach 79°N at both coasts, but are absent from Melville Bugt and also missing in parts of coastal parts of Northeast Greenland.

Six of the species have a continuous distribution in low arctic Greenland, while most of the rest occur north of 66° or 70°N. *Saxifraga hirculus* (Map 140) and *Eriophorum callitrix* (Map 160), are only known from Thule district on the west coast.

The type comprizes a heterogenous group concerning the biological distribution types. In this respect it is closely related to type 7 although the percentages of types are different. Most of the taxa (76%) belong to the circumpolar type.

Subtype 8a (Map Nos 144–160)
The species of this subtype have in addition to the main type isolated occurrences in the interior North Greenland. Fifteen of 17 species are missing from parts of coastal Northeast Greenland.

Five of the species have a more or less continuous distribution south of 79°N. *Ranunculus confervoides* (Map 152) is missing from most parts of Southeast Greenland. *Kobresia simpliciuscula* (Map 153) is not known from Southeast Greenland. *Carex rupestris* (Map 145), *Calamagrostis purpurascens* (Map 155), and *Erigeron eriocephalus* (Map 154) have their southern limit at 64°N, and *Draba alpina* (Map 144), *Carex marina* ssp. *pseudolagopina* (Map 157), *Ranunculus affinis* (Map 159), and *Ranunculus nivalis* (Map 149) have their southern limit at 66°N. *Carex atrofusca* (Map 156), *Puccinellia andersonii* (Map 158), and *Eriophorum callitrix* (Map 160) are not found south of 69°N.

Type 9 (Map No. 161)
This type includes only one species which has its northern limit south of 79°N on the west coast but is missing in Melville Bugt and has its northern limit south of 76°N on the east coast.

Arenaria humifusa is distributed in the main part of West Greenland and has also been found at a couple of sites in the area between 70°N and 74°N in East Greenland. It belongs to the middle arctic, amphi-Atlantic, type.

Type 10 (Map Nos 162–170)
This type includes species known in the study area only from Thule district. The Greenland distribution falls into two quite different distribution types. Four of the species have their Greenland distribution restricted to Thule district. The rest are low arctic, West Greenland species, which are known from most parts of the coast.

The type comprises a heterogenous group of taxa belonging to biological distribution types ranging from high arctic to boreal types. Most are western taxa (56%), the rest are circumpolar.

Type 11 (Map No. 171)
This type includes only *Geum rossii,* which is a new species to the flora of Greenland only found in one place in Lambert Land, Northeast Greenland.

It is arctic-alpine according to Porsild (1964) and mainly amphi-Beringian with few finds in High Arctic Canada and the one in Greenland.

Type 12 (Map Nos 172–177)
This type includes species which occur south of 76°N in both West and East Greenland.

All species have a continuous distribution in low arctic Greenland, and the holarctic distribution is exclusively low arctic with circumpolar, amphi-Atlantic, western and eastern elements.

Type 13 (Map Nos 178–182)
This type includes species occurring south of 76°N in West Greenland. All five species have a continuous distribution in low arctic Greenland.

Two third of the species are referred to the low arctic, widespread biological distribution type and the rest to the low arctic, oceanic type. Three of the five species are dwarf-shrubs. The main part (80%) has a amphi-Atlantic distribution. The rest are mainly found in the western Arctic.

Type 14 (Map Nos 183–217)
This type includes species occurring in the interior of Northeast Greenland south of 76°N. Of the 36 species 17 are known from the whole southern part of Greenland reaching their northern limit in West Greenland at app. 72°N and between 74°N and 76°N in East Greenland. Eight species are known from both West and East Greenland but are missing in South and Southeast Greenland. Of the remaining 10 species, which have a rather limited distribution in the central East Greenland, *Saxifraga nathorstii* (Map 216) and *Potentilla rubella* (Map 214), are endemic to Greenland.

Most species belong to the low arctic and boreal type (76%). Both oceanic and continental types are represented. The most important geographical distribution types are circumpolar, amphi-Atlantic, and eastern.

Subtype 14a (Map No. 218)
This subtype includes only *Polemonium boreale,* which is known from the interior as well as from the coastal parts of southern Northeast Greenland.

The species has a circumpolar distribution and is classified as arctic-continental.

5.3. Vegetation

Generally, northern Greenland is characterized by vast areas of almost barren landscape with very sparse vegetation. The vegetation cover is mostly less than a few per cent and the vegetated areas with a more or less continuous vegetation cover are only to be found in the lowland below 300 m. a.s.l. (Aastrup *et al.* 1986, Bay & Boertmann 1989). For North Greenland this lowland comprises app. 30% of the ice-free land. Most of the coastal areas can be termed polar desert (Babb & Bliss 1974, Bliss & Svoboda 1984, Alexandrova 1980, 1988) in having a total plant cover less than 5% (Fig. 11). Other areas, limited in extent, have considerable vegetation termed polar semi-desert and sedge-moss tundra (Bliss & Svoboda 1984).

Continental inland areas with very low precipitation have a desert-like appearance with no or few semi-permanent snow drifts persisting during summer, and the desiccation causes cracks in the surface of silty soils (Fig. 12). The major part (c. 95%) of these areas have a total vegetation cover less than 2–3 percent. The coastal areas have a thicker and prolonged snow-cover which can be a disadvantage to vegetation in cold summers when it melts very late, resulting in a reduction of the length of the growing season. Consequently, the vegetation cover here is as poor as in the continental parts, but the species diversity is lower as species with an ecological demand for summer warmth occur only in the sheltered inland. More or less continuous vegetation is definitely related to lowland areas with a sufficient water supply during most of the summer. Thus continuous vegetation is seen in snowbeds or on slopes below these and along streams, ponds, and lakes. Mosses constitute the main part of the vegetation cover in moist and wet habitats.

Fairly well-vegetated areas (cover: 25–30%) in northern Greenland have been located (Aastrup *et al.* 1986, Bay & Boertmann 1989) and by means of NOAA-satellite images covering the National Park in North and Northeast Greenland vegetated areas have been pointed out as important biological areas (Bay & Fredskild 1990). Similar high arctic oases with a dense vegetation cover are known from the Canadian Archipelago (Soper 1940, Svoboda & Freedman 1981, Bliss & Svoboda 1984, Muc *et al.* 1989, Henry *et al.* 1986).

5.3.1. Classification of the vegetation

No comprehensive publication about the vegetation in northern Greenland is available, only descriptions of minor areas are published. Bay (in Aastrup *et al.* 1986) classifies vegetation after analyzing the vegetation at seven localities from Peary Land in east to Hall Land in western North Greenland. A summary of the publications dealing with the vegetation of northern Greenland is given in Table 9.

The species composition and cover differ when moving along a climatic gradient, either from north to south or from coastal to inland areas. The description and classification are treated separately in the three regions: Northwest, North, and Northeast Greenland.

Seven main vegetation types are recognized throughout the area. The species diversity and composition vary with latitude. The general trend is an impoverishment in number of species, number and frequency of dwarf-shrubs, and the degree of cover when moving to the north and from inland areas to the coast. Vegetation analyses carried out by the author in part of the study area – mostly in North Greenland – together with the detailed descriptions form the basis of the classification. Degree of cover (%) was estimated for individual species of vascular plants, for total lichens, algae, bryophytes, and for unvegetated mineral substrate, and the vegetation types were defined on the basis of the floristic characteristics, physiognomy, and relation to topography and soil conditions.

5.3.2. The vegetation of Northwest Greenland

Sørensen (1943) was the first to give a description of the vegetation in Melville Bugt based on notes from Salomonsen. Tugtuligssuaq is the only area in Melville Bugt with detailed vegetation descriptions (Fredskild & Bay 1980, Fredskild *et al.* 1979, Fredskild 1985). Fredskild (1985) and Bay (Fredskild & Bay 1989) give detailed descriptions of the vegetation of minor areas in Thule district.

1) Aquatic habitats

Ponds and lakes are very poor in species of vascular plants. Mosses are occasionally seen as a carpet down to a depth of 5 metres. *Ranunculus hyperboreus* is the only aquatic plant found in Melville Bugt, whereas few other species are rare in Thule district: *Pleuropogon sabinei, Ranunculus confervoides, Hippuris vulgaris,* and *Eleocharis acicularis. Hippuris vulgaris* formerly occurred in a lake at Tugtuligssuaq as proved by pollen analyses (Fredskild & Bay 1980, Fredskild 1985).

2) Fen and grassland

In Melville Bugt only one type of fen is found. The dominating species are *Eriophorum scheuchzeri, E. triste,* and *Carex stans* and the subdominants are *Juncus biglumis, J. castaneus,* and *Eriophorum angustifolium.* Being mainly topographically conditioned these fens cover only small areas contrary to Thule district where large lowland areas are covered by fens. In the northern part of the region additional species occur in the fens: *Arctagrostis latifolia, Dupontia fisheri, Carex saxatilis,* and *Deschampsia brevifolia.* On richer soil another type which in addition includes *Carex atrofusca, Kobresia simpliciuscula,* and *Juncus triglumis* is found. *Tofieldia coccinea* grows on hummocks in this type of vegetation.

3) Dwarf-shrub heath

The number of dwarf-shrubs decreases northwards in Melville Bugt. The low arctic species *Phyllodoce coerulea, Loiseleuria procumbens, Betula nana,* and *Harrimanella hypnoides* reach their northern limit at Kangerdluarssuk (74°18′N), and north of here the dwarf-shrub heaths are poor in species, *Salix arctica, Vaccinium uli-*

Fig. 11. Vast areas in North Greenland are almost without vegetation, this being restricted mainly to areas below snow drifts, which supply the plants with water during the growing season.

Fig. 12. Desiccation cracks frequently occur in continental parts of North Greenland where the precipitation is very low.

Fig. 13. *Ranunculus hyperboreus* and *Pleuropogon sabinei* are the only frequent species of vascular plants occurring in lakes and ponds. They prefer the continental part of northern Greenland.

Fig. 14. Fen vegetation has a low species diversity. *Carex stans* and *Eriophorum scheuchzeri* occur in the most wet parts surrounded by a zone dominated by *Eriophorum triste.*

Fig. 15. Hummocky dwarf-shrub heath dominated by *Salix arctica* is the most common heath type north of 80°N. *Oxyria digyna, Polygonum viviparum, Papaver radicatum,* and *Festuca hyperborea* are frequent in this type.

Fig. 16. Late snowbed vegetation with *Phippsia algida, Saxifraga hyperborea,* and *S. tenuis* is common in the outer coast districts.

Fig. 17. Manured habitats at bird perches and lemming borrows are dominated by *Puccinellia angustata, Papaver radicatum, Draba arctica,* and *Alopecurus alpinus.*

Fig. 18. Luxuriant dwarf-shrub heath is found in the southern part of Northeast Greenland. *Vaccinium uliginosum, Cassiope tetragona, Empetrum nigrum,* and *Salix arctica* dominate on south-facing slopes in the lowland.

Fig. 19. South-facing herb-slopes in the northern part of Northeast Greenland. The lush vegetation is dominated by *Oxyria digyna, Polygonum viviparum, Arnica angustifolia, Stellaria edwardsii,* and *Salix arctica.*

Fig. 20. *Salix arctica* snowbed occurring on sloping terrain is a distinct type of vegetation in Northeast Greenland.

ginosum, and *Cassiope tetragona* being the only common species. The low arctic species are of minor importance north of Laksefjord (72°30′N) as only few suitable sites occur. Copses of *Salix glauca* do not occur in the region and the species is rare on Tugtuligssuaq (75°25′N). *Vaccinium uliginosum* and *Dryas integrifolia* dominate on the early snow-free south-facing slopes whereas *Cassiope tetragona* is frequent on level ground where the snow melts later. Common herbs in the heaths are *Luzula confusa, Pedicularis hirsuta,* and *Carex bigelowii.* In Thule district a lush type on moist silt dominated by *Vaccinium, Cassiope, Salix arctica,* and *Rhododendron lapponicum* occurs. This heath type mostly covers only few square metres on plain ground. *Cassiope* is the only dwarf-shrub dominating in larger areas. Common herbs are *Carex bigelowii, Oxyria digyna, Polygonum viviparum, Pyrola grandiflora,* and *Cerastium arcticum. Dryas integrifolia* is dominating on dry soils often together with *Carex nardina, Kobresia myosuroides,* and rarely, *Carex supina* and *C. rupestris.*

4. Snowbed and herb-slope

Long lasting snowbeds are dominating on level ground and on the lower part of the slopes. *Anthelia juratzkana*-snowbeds usually without phanerogams are common in late snow-free areas, where the growing season is reduced to four weeks. *Luzula arctica* – often sterile – is one of the few species of vascular plants able to establish in this community. *Salix herbacea* dominates in early snow-free snowbeds above the *Cassiope tetragona* dominated heath.

Herb-slopes are rare in Melville Bugt. On Tugtuligssuaq this type is found in one place only occupying few square metres on a moist south-facing slope. *Oxyria digyna, Ranunculus pygmaeus, Antennaria canescens, Carex bigelowii,* and *Salix herbacea* were growing in a

Table 9. Summary of the publications concerning vegetation in northern Greenland.

Area	Region	Latitude	Remarks	References
Tugtuligssuaq	1	75°25′	Detailed descriptions	Fredskild *et al.* 1979 Fredskild & Bay 1980 Fredskild 1985
NW Gr.	1,2	72–79°	Brief descrip.	Bay 1983
Melville Bugt	1	73–76°	Brief discrip. + classification	Sørensen 1943
Melville Bugt + Thule dist.	1,2	74–77°	Detailed descrip. + classification	Jakobsen 1950
Thule distr.	2	76–78°	Detailed descrip.	Fredskild & Bay 1989
Qaanaaq	2	77°30′	Detailed descrip. of lichen vegta.	Hansen 1987 Hansen 1989
Qeqertat	2	77°30′	Brief descrip.	Fredskild 1985
Kap Inglefield	2	78°30′	Brief descrip.	Blake *et al.* 1992, Tedrow 1970
Coastal N Grl.	3	80–83°	Short note	Ostenfeld 1925a
Coastal N Grl.	3	80–83°	Notes and classification	Ostenfeld 1923b
Cass Fj, Washington Ld.	4	80°	Brief descrip. + classification	Bonneval 1977
North Grl.	3,4	80–83°	Detailed descrip. analyses + clas.	Aastrup *et al.* 1986
Kap København	4	82°30′	Brief descrip. of moss vegetation	Mogensen 1985
Peary Land	4	82°30′	Brief description	Fredskild 1966b, 1973
Peary Land	3,4	82°30′	Brief descrip.	Holmen 1957
Peary Land	3,4	82–83°	Classification + description	Holmen 1955
Nansen Land	4	83°	Detailed descrip., biomass analyses	Fredskild *et al.* 1992, Klein & Bay in prep.
Kap København	4	82°30′	Detailed descrip., biomass analyses	Fredsklid & Bay 1988, Klein & Bay 1990
Kap København	4	82°30′	Brief description Detailed descrip.	Bennike 1990, Fredskild *et al.* 1992
Blåsø	4	80°	Detailed descrip., biomass analyses	Fredskild & Bay 1988, Klein & Bay 1990
Brønlund Fjord	4	82°30′	Remarks + descrip.	Fredskild pers. comm.
Brønlund Fjord	4	82°30′	Analyses	Fredskild pers. comm.
Prinsesse Ingeborg Halvø	3	81°30′	Detailed descrip. + analyses	Fredskild & Bay 1987
Kilen	3	81°	Short description	Mølgaard 1985
Skærfjorden	6	77°30′	Detailed descrip. + classification	Cabot *et al.* 1988
NE Grl.	5,6 7,8	72–78°	Brief descrip.	Bay & Boertmann 1989
NE Grl.	5,6 7,8	74–79°	Detailed descrip. + classification	Bay 1991, Bay & Fredskild 1990, 1991
NE Grl.	5,6 7,8	74°30′–79	Detailed descrip. + classification	Seidenfaden & Sørensen 1937
NE Grl.	7,8	72–75°	Description + classification	Oosting 1948
NE Grl.	5,6	77°	Description	Lundager 1912
NE Grl.	7,8	73–74°30′	Notes	Seidenfaden 1930
NE Grl.	7,8	73–74°30′	Notes	Seidenfaden 1931
NE Grl.	8	74–75°	Description and classification	Schwarzenbach 1961
Zackenberg	8	74°30′	Description and classification	Fredskild & Mogensen 1991, Fredskild *et al.* 1992

moist carpet of moss. In Thule district herb-slopes, or rather, early snow-free snowbeds occur in sheltered sites with a warm microclimate. The characteristic species are *Oxyria digyna, Poa arctica, Minuartia biflora, Silene acaulis, Potentilla hyparctica, Salix arctica,* and besides *Arnica angustifolia* in more dry places. The cover of vascular plants often exceeds 50%.

5. Fell-field and vegetation on solifluction soils

Fell-field includes wind-swept ridges, barren ground, and solifluction ground all with a cover of vascular plants less than a few per cent. The most common species on dry ground are *Luzula confusa, Carex nardina, Poa glauca, Cardamine bellidifolia, Draba nivalis,* and *Antennaria ekmaniana.* Occasionally *Diapensia lapponica, Carex glacialis, Saxifraga tricuspidata,* and *Campanula uniflora* occur. Habitats dominated by *Kobresia myosuroides* and with *Carex rupestris, C. supina,* and *Potentilla hookeriana* might be referred to as a kind of steppe. *Carex rupestris* is very rare in Melville Bugt in contrast to Thule district, where it is one of the characteristic species on dry and wind-swept habitats.

Transitions from late snowbed vegetation to vegetation on solifluction soil occur frequently on Tugtuligssuaq. In places where the soil is too unstable, the *Anthelia juratzkana*-snowbeds are replaced by a very open vegetation with a few *Saxifraga tenuis, S. hyperborea,* and *Sagina intermedia* and mostly without mosses.

6. Saltmarsh and sandy-beach vegetation
Judging from the distribution of *Puccinellia phryganodes* and *Stellaria humifusa* saltmarshes are widespread in the region although here they are always fragmentary. In addition to these, *Cochlearia groenlandica, Carex glareosa,* and *C. ursina* grow in the outer saltmarsh, *Saxifraga rivularis, Sagina intermedia, Koenigia islandica,* and *Phippsia algida* occur in the inner part.

Sandy-beach vegetation with *Honckenya peploides* is only found occasionally in Thule district.

7. Manured habitats
The many bird cliffs in Thule district give rise to a lush vegetation mostly dominated by *Alopecurus alpinus.* The plants are up to half a metre and a vigorous peat formation is present. Accompanying species are *Poa arctica, Hierochloë alpina,* and *Stellaria longipes.* The species diversity is extremely low. Bird perches, which are a characteristic element in the landscape of North Greenland, have only been found once in Thule district.

5.3.3. The vegetation of North Greenland

A classification of the phanerogam vegetation in North Greenland is given by Bay in Aastrup *et al.* (1986) and the phanerogam and bryophyte vegetation of Peary Land is given by Holmen (1955, 1957). Notes of the vegetation from local areas are published by Bay in Fredskild *et al.* (1987, 1992) and in Fredskild & Bay (1988), in addition to Bonneval (1977), Fredskild (1966b, 1973), Mogensen (1985), and Mølgaard (1985).

1) Aquatic habitats
Pleuropogon sabinei and *Ranunculus hyperboreus* are the only species of vascular plants occurring commonly in ponds. *Ranunculus hyperboreus* is more rare than *Pleuropogon* and seems to prefer the continental inland. *Hippuris vulgaris* and *Ranunculus confervoides* have only been recorded a few times mostly from the interior of Peary Land (Fig. 13).

2) Fen and grassland
The wetland vegetation includes *Carex stans-Eriophorum scheuchzeri* fens, *Eriophorum triste* fens, and a graminoid vegetation with a higher species diversity. *Carex stans* fens occur on silty, level ground along ponds and streams or below snowbeds. *Carex stans* forms pure stands in the most wet places. *Eriophorum triste* dominates on moist, sloping ground with hummocks. These fens often replace the *Carex stans* fens in the less wet places (Fig. 14).

Grassland has been found in few sheltered places on moist, sloping ground and in depressions with a constant water supply but less moist than in fens. The dominant species are the graminoids *Alopecurus alpinus, Arctagrostis latifolia, Deschampsia brevifolia, Carex stans, Eriophorum triste,* and the herbs *Ranunculus sulphureus, Draba lactea, Cardamine bellidifolia, Melandrium apetalum,* and *Polygonum viviparum.* Occasionally, *Pleuropogon sabinei* occurs in the most wet part of the grassland. The species diversity is one of the highest of the North Greenland vegetation types. The degree of cover is similar to that of the fens. Phanerogams cover 5–10% and mosses 50–100%. In contrast to the *Carex stans* fens *Salix arctica* occurs on the hummocks.

3) Dwarf-shrub heath
Ostenfeld (1923b) states that no real heath occurs in the area visited by Wulff. This is true for the outer coast, but in the sheltered inland areas a closed vegetation dominated by one of the three common dwarf-shrubs may occur in minor areas. The heath can be classified according to the dominant dwarf-shrub species, which are depending on the water content in the soil and the duration of the snow-cover. *Salix arctica* dominates in dwarf-shrub heaths on hummocky ground (Fig. 15). This is the most widespread type, with a cover of only a few per cent of vascular plants against 25–50% of mosses. Common herbs are *Papaver radicatum, Festuca hyperborea, Oxyria digyna, Saxifraga cernua, Stellaria crassipes,* and *Luzula arctica. Alopecurus alpinus* occurs in the most moist places. A species diversity of 24 is recorded in Peary Land. Patchy *Cassiope tetragona* heaths occur in snow-covered depressions on level ground or on the south-exposed hillsides, where there is lee for the prevailing winds from northeast. *Cassiope tetragona* covers up to 50% and a moss carpet usually covering more than 50% is common. *Dryas integrifolia* and hybrids between *D. integrifolia* and *D. octopetala* dominate the heaths on dry ground. Such are found on windswept, level ground and as a zone above the *Cassiope* heath on sloping terrain, and is almost snow-free during winter. Accompanying species are *Cerastium arcticum, Pedicularis hirsuta, Kobresia myosuroides,* and *Stellaria crassipes.* The species diversity is low; *Dryas integrifolia* covers less than 20%, and mosses are rare.

4) Snowbed and herb-slope
Snowbed vegetation is common in connection with accumulated snow on sheltered hillsides. The vegetation can be divided according to the duration of the snow cover. The early snow-free type is the richest with a high species diversity. Characteristic species are *Taraxacum arcticum, T. pumilum, Saxifraga cernua, Oxyria digyna, Festuca hyperborea,* and *Alopecurus alpinus. Salix arctica* is the only woody plant in this type. Typical herb-slope vegetation is missing in North Greenland. However, a few square metres in the early snowbed could rarely be designated "high arctic herb-slope". Late snow-free snowbed vegetation has a lower diversity of

phanerogams but a more dense moss-cover (Fig. 16). *Saxifraga hyperborea, S. tenuis*, and *Phippsia algida* are characteristic for this type which lacks woody plants.

5) Fell-field and vegetation on solifluction soils
In dry places in the lowland fell-field is widespread and vast areas in the highland are dominated by this type. Wulff states (Ostenfeld 1923b) in his diary that all the vegetation formations in North Greenland can be designated fell-field. Ostenfeld (1923b) modifies the statement of Wulff by his interpretation stating that the dominant formation is fell-field but other formations also exist, yet are of inconsiderable extent only, and occur merely as patches in the fell-field where edaphic conditions allow for their existence. Fell-field is characterized by *Poa abbreviata, Kobresia myosuroides, Poa glauca, Draba subcapitata,* and *Minuartia rubella.* The total plant cover, less than 1%, is low due to absence of mosses.

On wet to moist solifluction soil a vegetation type with a correspondingly low species diversity and degree of cover is found. *Cerastium regelii, Minuartia rossii, Phippsia algida,* and *Colpodium vahlianum* characterize this type on relatively stable soil with an organic crust.

6) Saltmarsh and sandy-beach vegetation
Most of the species characteristic of saltmarsh have their northern limit south of North Greenland. Consequently, saltmarshes in North Greenland are very poor in species. The long duration of the sea-ice cover and the erosion of the ice on the shores leave only few suitable places for this vegetation type. *Cochlearia groenlandica* and occasionally *Puccinellia phryganodes* and *Phippsia algida* occur.

7) Manured habitats
A conspicuous element in the North Greenland landscape is the bird perches which appear in connection with stones slightly elevated above the surroundings. A lush vegetation is formed here and besides at places with lemming borrows and muskoxen carcasses where an extra supply of nutrition is released. The frequent species are *Puccinellia angustata, Papaver radicatum, Draba arctica, Draba subcapitata,* and *Alopecurus alpinus.* They occur in the surroundings but are much lusher on the manured places (Fig. 17).

5.3.4. The vegetation of Northeast Greenland

The largest material on vegetation of the study area is from this region. Seidenfaden & Sørensen (1937) and Bay & Fredskild (1990, 1991) give a comprehensive description of the vegetation in the area between 74°30′N and 79°N and 75°–79°10′N, respectively. Several botanists deal with the vegetation of local areas: Germania Land (Lundager 1912), Skærfjorden (Cabot *et al.* 1988), Sabine Ø-Wollaston Forland (Hartz & Kruuse 1911), Clavering Ø-Wollaston Forland (Gelting 1937), and the nunatak zone between 74° and 75°N (Schwarzenbach 1961).

This region has a marked vegetational border in the northern part of Hochstetter Forland. The lowlands south of here have a continuous cover of vegetation mostly because of a dense moss carpet. North of Bessel Fjord vast areas in the lowland and all highland areas have a low degree of cover and the more luxuriantly vegetated places are rare and have a local occurrence with a limited distribution.

1) Aquatic habitats
The most common aquatic species is *Pleuropogon sabinei,* which forms dense stands along the edge of ponds. *Ranunculus hyperboreus* is common in contrast to *Hippuris vulgaris* and *Ranunculus confervoides.* Like *Potamogeton filiformis* the latter is restricted to inland areas.

2) Fen and grassland
The most widespread vegetation type in the wetlands is a *Carex stans-Eriophorum scheuchzeri* dominated fen with *Carex saxatilis* and *Juncus biglumis.* It occurs on silty soils which have a persisting high water level during summer. In this respect it differs from grassland vegetation which dries out at the end of the growing season. The species diversity is low in contrast to a type of fen which occurs locally on rich soils. Characteristic species of this type are *Carex atrofusca, Kobresia simpliciuscula, Juncus triglumis,* and *Eriophorum callitrix.* Dry grassland is dominant in the interior parts and the characteristic species are *Kobresia myosuroides* and *Carex rupestris.* Grassland is the most widespread of the vegetation types with a continuous cover of plants constituting important habitats for the herbivor animals.

3) Dwarf-shrub heath
The southern parts of Northeast Greenland has a higher number of woody plant species. *Betula nana* forms a type of mixed dwarf-shrub heath with *Vaccinium uliginosum, Cassiope tetragona, Empetrum nigrum,* and *Salix arctica* on south and west-facing slopes in the lowland in the southernmost part of the region (Fig. 18). It dominates in the inland and the frequency decreases towards the coast. The low arctic herbs *Pedicularis lapponica, Tofieldia pusilla,* and *Campanula gieseckiana* have been found a few times in this type.

The *Cassiope tetragona* heath, which is the most widespread type in the region, forms a conspicuous zone on the lower parts of the lee sides of hills and in depressions. Seidenfaden & Sørensen (1937) state that there is a connection between this vegetation type and the soil as it occurs mostly on sandy-gravelly outwashed soils. Often the *Cassiope tetragona* cover is very closed (75–100%) only leaving space for few more species. *Huperzia selago* is exclusively associated with this heath type. Other species are *Pedicularis hirsuta, Oxyria digyna, Silene acaulis,* and occasionally *Tofieldia coccinea.*

Salix arctica, which has a wide ecological tolerance,

never forms true heath vegetations. It occurs along an ecological gradient from wind-exposed ridges over *Cassiope, Dryas,* and *Vaccinium uliginosum* dominated heaths to late snowbed vegetations. *Vaccinium* heath occurs as a zone above the *Cassiope* vegetation on sloping terrain and is early snow-free. The type is only frequent in the southern parts. *Dryas octopetala* mixed with hybrids dominated heaths are mostly occurring on level or south-facing terrain throughout the region. *Kobresia myosuroides* and *Carex rupestris* are always associated with *Carex nardina* and *Poa glauca. Arctostaphylos alpina* is an element in this type in the most southern part of the area, *Rhododendron lapponicum* locally occurs in the moist, lush mixed heath and in *Dryas* heath. *Empetrum nigrum* is rare and never forms true heaths.

4) Snowbed and herb-slope
Snowbeds with a high species diversity are found locally on sheltered slopes, especially in the southern part. Species exclusively associated with this early snow-free snowbed are *Taraxacum arcticum, Ranunculus pygmaeus, R. nivalis, Trisetum spicatum, Minuartia biflora,* and occasionally *Salix herbacea.* The degree of cover is less than 10% but locally *Salix arctica* is more dominating resulting in a larger cover. In the most lush places true herb-slopes occur. In addition to the species mentioned *Erigeron humilis, E. eriocephalus, Euphrasia frigida,* and *Gentiana tenella* occur. *Arnica angustifolia, Draba fladnizensis, Campanula uniflora,* and *Festuca vivipara* are present in the more dry parts of the herb-slopes. In the northern part the herb-slopes have a lower species diversity and are without the low arctic species (Fig. 19). Goodwillie gives a list of 15 species (Cabot *et al.* 1988), but judging from the knowledge of the ecology of some of these species it is from a very wet herb-slope with transitions to wetland vegetation.

In the northern parts a *Salix arctica* dominated snowbed vegetation is common on sloping terrain (Fig. 20). It occupies the zone above the *Cassiope* heath and substitutes the *Vaccinium uliginosum* zone known from the southern part. This type is widespread in Jameson Land (70–71°N) in East Greenland (Bay & Holt 1986), and has a snow-cover of a larger depth and longer duration. Gelting (1937) measured 100 cm of snow in late winter on this vegetation type whereas the snow depth varied between 110 and 230 cm in Jameson Land (Bay & Holt 1986).

5) Fell-field and vegetation on solifluction soils
Fell-field dominates in the highland and on flat, wind-swept plains, wind-exposed slopes and rocks with an unfavourable exposure. The snow-cover is missing or very thin. Gelting (1937) has recorded a cover of 0–10 cm in an open *Dryas octopetala* vegetation. The snow is redistributed several times during winter resulting in frequent exposure of the plants easily visible on the asymmetric wind-eroded growth form of the plants. *Carex nardina, Poa abbreviata, Potentilla rubricaulis, Papaver radicatum,* and *Minuartia rubella* are characteristic of this plant community. *Dryas octopetala* and *Salix arctica* are the only woody plants occasionally occurring here.

On constantly wet, unstable soil especially in the outer coast region a vegetation with a cover of phanerogams as small as in the fell-field is found. In spite of a large species diversity the cover of phanerogams does not exceed 1%, whereas the cryptogam cover of lichens and algae is large – often 50–100%. In this community, which appears as a black crust below snowdrifts, *Sagina intermedia, Phippsia algida, Saxifraga foliolosa, S. tenuis, Ranunculus sulphureus, R. glacialis, Draba lactea,* and *D. adamsii* are found. This type is very common north of Dove Bugt (76°N).

6) Saltmarsh and sandy-beach vegetation
The fragmentary saltmarshes of the region occupying only very few square metres on suitable places at sheltered coasts are dominated by: *Carex subspathacea, C. ursina,* and *Stellaria humifusa.* Sandy-beach vegetation is very rare. *Honckenya peploides* has only been found a few times, mostly in the interior parts of the region.

7) Manured habitats
Bird perches are very rare in Northeast Greenland. Only in a few places species from dry plant communities are growing here. Like in North Greenland the most common species is *Puccinellia angustata.* Moreover *Draba arctogena* and *Potentilla hookeriana* are found.

5.4. Vegetation mapping

In connection with the biological mapping of Northeast Greenland NOAA-satellite images covering the whole National Park in North and Northeast Greenland have been used. Maps showing the density of the vegetation from Scoresby Sund to Humboldt Gletscher have been worked out. The two maps (scale c. 1:2, 380000) covering the eastern part of the study area are presented in Figs 21–22. Recording of vegetation from satellite images are based on the fact that different kind of surfaces have different reflection. Chlorophyll and other pigments in the plants absorb a part of the influx of light resulting in a reflection different to unvegetated areas. The intensity of the reflection from the vegetation expresses the degree of cover. The single picture element, pixel, covering one square kilometre, represents the average reflection from the ground within two spectral bands: 560–680 nm and 725–1100 nm, corresponding to green and nearinfrared light.

The maps show the vegetation at its maximum development which has been determined to the beginning of August comparing a number of images taken throughout the growing season. The Normalized Difference Vegetation Index (NDVI) consists of 6 categories with values from −0,5 to 0,25 corresponding to a degree of cover of 0% to app. 30% (Hansen & Søgaard 1989).

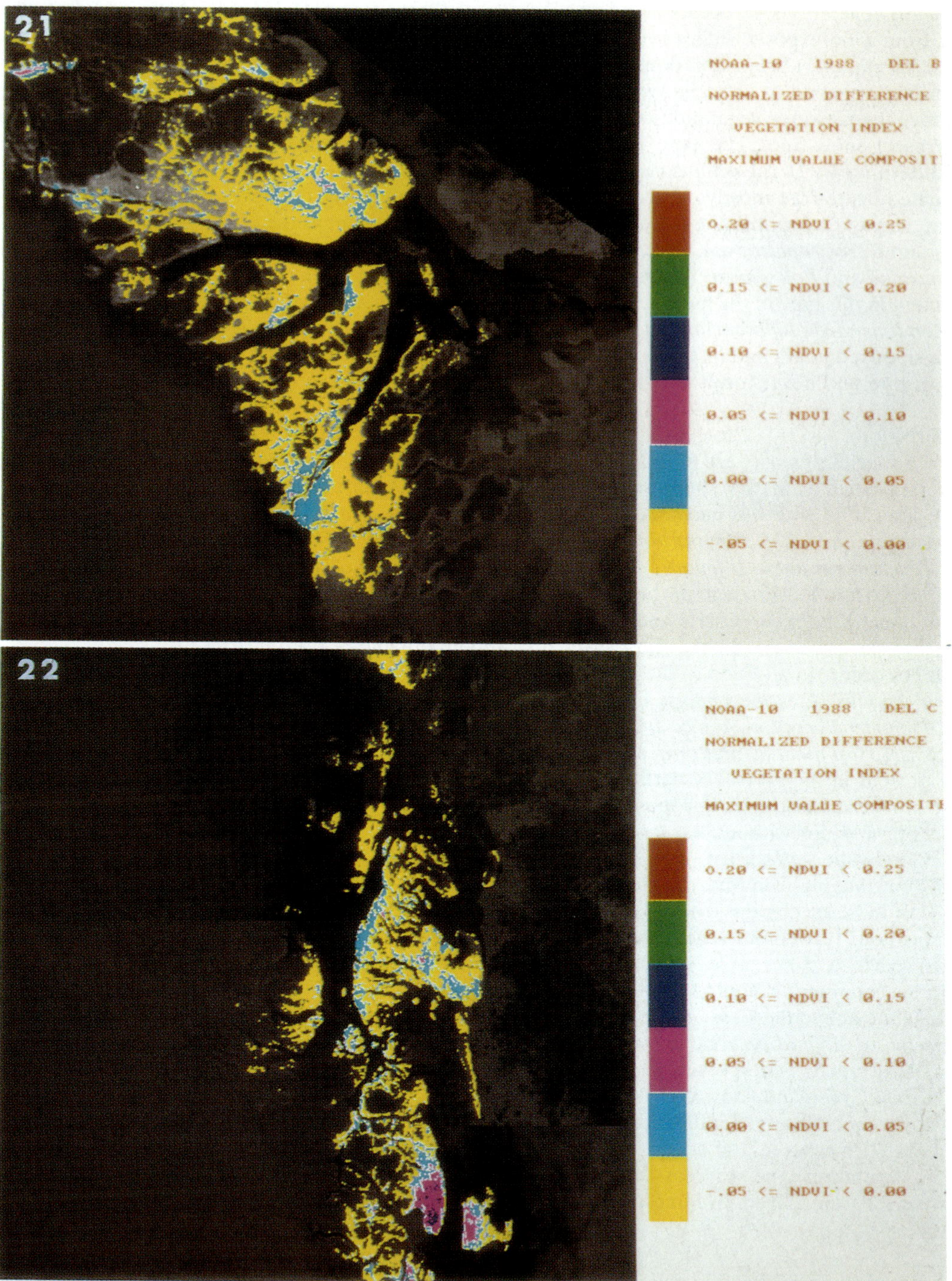

Fig. 21. Vegetation map of eastern North Greenland based on NOAA-satelitte images showing the degree of cover of the vegetation. Low NDVI values represent areas with a poor vegetation cover, high values luxuriant areas.

Fig. 22. Vegetation map of North-east Greenland based on NOAA-satelitte images showing the degree of cover of the vegetation. Low NDVI values represent areas with a poor vegetation cover, high values luxuriant areas.

Judging from the satellite images the richest areas are the lowland east of Zackenberg and Albrechts Bugt in Wollaston Forland and the southwestern part of Kuhn Ø. North of here no areas have a NDVI higher than 0,15. Most of the remaining lowland south of 76°N is referred to the two middle categories. These areas, having a cover of app. 10–20%, have their main distribution on Hochstetter Forland, Shannon, the west side of Kuhn Ø, northern Wollaston Forland, and the west side of Clavering Ø. The two categories comprizing areas with a cover of less than c. 10% are mostly found in the interior upland.

North of Hochstetter Forland no lush areas with a cover larger than c. 15% are found in Northeast Greenland. Two areas in Germania Land and few in North Greenland have been referred to the category with a NDVI between 0,05 and 0,10 which corresponds to a cover of c. 8–15%. The main part of the vegetated areas between 76° and 80°N have a cover of less than c. 8%. These areas are mostly found in the inland from the head of Bessel Fjord through A. S. Jensen Land, Lindhard Ø, Daniel Bruun Land, Okselandet to Nordmarken (Fig. 22) and the central Germania Land. These areas cover 24% of the ice-free land.

Brainard Sund in Nansen Land is the largest luxuriant area in North Greenland (Fredskild *et al.* 1992, Klein & Bay in prep.). Although the NOAA-satellite images of western North Greenland are of a low quality, it can be concluded that west of Nansen Land vegetated areas are small and very scattered, and have a degree of cover always less than 8%. The central inland in Peary Land and areas at the head of Danmark Fjord are the only areas mapped of this category. Such areas can be considered high arctic oases, which are very important biological areas with a high diversity and density of plants and animals.

The floristic boundary at Bessel Fjord (76°N) in East Greenland is coincident with a distinct vegetational boundary judging from the NOAA-satellite images and the aerial survey in 1988. Thus the lush vegetation with a cover larger than 25% in Wollaston Forland and the west side of Kuhn Ø as well as the well-vegetated areas on Hochstetter Forland and the west side of Shannon with a cover of 10–20% are replaced by a more sparse vegetation in the main part of the lowlands north of Bessel Fjord.

6. Discussion and conclusion

6.1. Delimitation of floristic provinces and districts

The floristic province North Greenland is defined phytogeographically as the northernmost part of Greenland north of Humboldt Gletscher (79°N) and Lambert Land (79°30′N) in West and East Greenland, respectively (Simmons 1909, Böcher *et al.* 1959) (Fig. 23). The present study has confirmed this delimitation. 40 species have their northern limit just south of 79°N in Northwest Greenland. The corresponding number in Northeast Greenland is 32 (Fig. 8). Based on the much larger material now at hand the North Greenland province has been divided into two districts: 1) the interior North Greenland (N_i), and 2) the coastal North Greenland (N_c) according to the distribution patterns of selected species (Fig. 9).

Holmen (1955) divides the vegetation types of Peary Land and adjacent areas into two groups, corresponding to two types of climatic conditions. Surprisingly, he includes the most continental areas at the head of Independence Fjord in the category that comprises coastal areas (Fell-Field community). Further, Holmen includes the extreme coastal areas at Prinsesse Ingeborg Halvø and the islands west of here in the other type (High Arctic Desert community) which otherwise consist of continental areas in the interior Peary Land. Consequently, his division is only partly in agreement with the proposed delimitation. The delimitation of the polar desert zone in North Greenland proposed by Funder & Abrahamsen (1988) agrees for the eastern part with the present delimitation of the coastal floristic district of North Greenland as proposed above.

The coastal district in North Greenland includes the most coastal parts of North Greenland and is within the polar desert zone defined by Alexandrova (1980, 1988), Elvebakk (1985), and Young (1971). According to these authors only the coastal areas of northern Peary Land westwards to Hall Land in western North Greenland are included in this zone. However, the present study shows that this boundary should be altered as indicated by Funder & Abrahamsen (1988), including the areas around Flade Isblink to the south to app. 80°N in eastern North Greenland. The delimitation in western North Greenland is more uncertain as a smaller number of collections are available. According to Edlund (1983) who classified the Canadian High Arctic these areas can be included in her bioclimatic zone 1.

The extreme coastal areas are characterized by low species diversity without plants of the genera *Carex* and *Eriophorum*. Edlund (1983) stated that only 35 species of herbs are found in these extreme areas in arctic Canada whereas 55 spp. are found in the equivalent zone in North Greenland. Dwarf-shrubs are almost absent, and only occasionally non-flowering prostrate specimens of *Salix arctica* are seen. Total plant cover is less than 5% and large areas are unvegetated. Physically, the areas within the zone is characterized by very low summer temperatures, averagely 2–2,5° C in July, and low precipitation.

The border at 76°N between the two districts in Northwest Greenland is confirmed too. Thus 64 species have their southern limit and 12 species their northern limit here. With 76 distribution limits this is the most distinct border in the study area. It differs from the other borders in having a larger number of southern

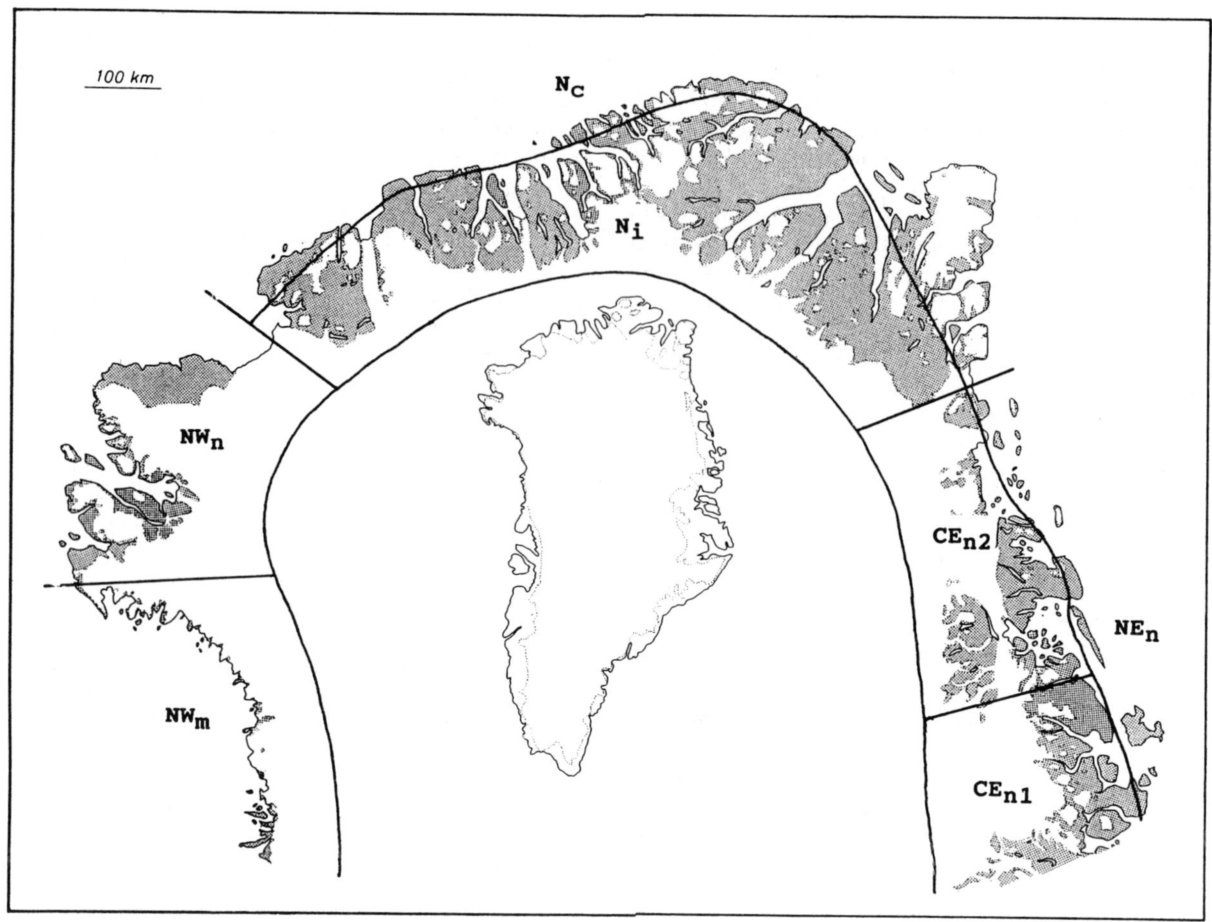

Fig. 23. Revised delimitation of the floristic provinces and districts in northern Greenland. Three new districts are proposed.

limits compared to northern limits (Fig. 8). A combination of climatic, edaphistic and historical factors may explain this, whereas the other distribution limits are mostly considered as reflecting either climatic or historical events.

Melville Bugt can be considered exclusively a coastal area as the ice-free areas are very small and no inland areas occur. In addition there is a lithological border in the northern part of Melville Bugt, where acid gneisses are replaced by neutral sandstone.

Species usually considered as preferring continental areas are not restricted to the eastern part of Inglefield Bredning in Thule district, which is the only area that could be considered distinctly continental in this district. This fact can be explained by local climatic conditions along with different geological conditions. Acid gneiss is dominating in the eastern part whereas the more or less alkaline sedimentary bedrock occurs in the western part. Consequently, Thule district has not been divided into a coastal and an inland district.

Böcher *et al.* (1959, 1978) divide Northeast Greenland into a coastal (NE) and a continental (CE) floristic province each divided into three districts: northern (n), middle (m), and southern (s). The present study area comprises the northern districts of NE and CE plus the northernmost part of the middle district in NE (Fig. 24). The intensive field work in this region in 1989 and 1990 has given a detailed knowledge of the local distribution of the species. Mapping of selected low and high arctic species reveals that the low arctic species (e.g. *Betula nana* (Map 173), *Rumex acetosella* (Map 196), and *Campanula gieseckiana* (Map 174) prefer the sheltered inland areas when approaching their northern limit at app. 76°N in the study area. Several of these species are only known from the interior parts of C. H. Ostenfeld Land and Hochstetter Forland. On the contrary, some high arctic species (e.g. *Saxifraga platysepala* (Map 62), *Cerastium regelii* (Map 78), and *Taraxacum pumilum* (Map 81)) have a nearly complementary distribution in the southern part of their range, only occurring in the coastal areas (Bay & Fredskild 1990). The northern district (CEn) in the continental floristic province includes the nunatak area Dronning Louise Land. No evidence for including this fairly high arctic area in a district, which also includes the head of the fjords as far south as 73°N, has been found except for the boreal, aquatic species *Potamogenton filiformis* (Map 129) found once. Consequently, this district is divided into two districts: CE_{n1} and CE_{n2} (Fig. 23). The southern district includes Clavering Ø, Wollaston Forland, A. P. Olsen Land, Th. Thomsen Land, C. H. Ostenfeld Land, and Hochstetter Forland plus continental areas south of 74°N, originally included in CE_n. The northern district includes Adolf S. Jensen Land, Dronning Louise Land, all islands in Dove Bugt (excl. Store Koldewey), central parts of Germania Land, Søndermarken, and Nordmarken. The border is placed by Bessel Fjord at app. 76°N thus, it is proposed to divide CE into 4 districts.

Surprisingly, the delimitation by Böcher *et al.* (1959) of coastal and continental areas in the floristic provinces of Northeast Greenland does not follow Seidenfaden & Sørensen (1937) even though no additional material from the southern part of the area had been collected in the intervening period. The biggest differences between these proposals are north of Bessel Fjord. The delimitations by the above mentioned authors are shown in Fig. 24. The field work in 1989 and 1990 showed that the division by Seidenfaden and Sørensen is more in agreement with the one proposed in this paper (Fig. 23). The border proposed by Seidenfaden and Sørensen is more easterly than that of Böcher *et al.* (1959) including only the coastal parts of Hochstetter Forland, Adolf S. Jensen Land, Store Koldewey, and the eastern part of Germania Land, east of a line from Hvalrosodden to the western part of Flade Bugt, and further to the eastern part of Skærfjorden in the outer coast area.

A new delimitation of the coastal and continental areas in Northeast Greenland is shown in Fig. 25. The distribution of 39 selected middle and low arctic species almost exclusively with an occurrence in the continental part of the area is used for the delimitation. The species mapped belong to NGDT 5, 5a, 7, 8a, 12, and 14. The proposed delimitation differs from the one by Seidenfaden and Sørensen with respect to include minor areas in the coastal districts, now only including eastern parts of Wollaston Forland, Shannon, Store Koldewey, eastern Germania Land, northern part of Stormlandet, the islands east of the 20th western longitude, and eastern part of Lambert Land.

The border between NE_m and NE_n proposed by Seidenfaden and Sørensen (1937) and slightly modified by Böcher *et al.* (1959) (Fig. 24) cannot be confirmed, as no species has a distribution limit in the middle part of coastal Wollaston Forland. The only two species with a distribution limit in this region are *Taraxacum pumilum*

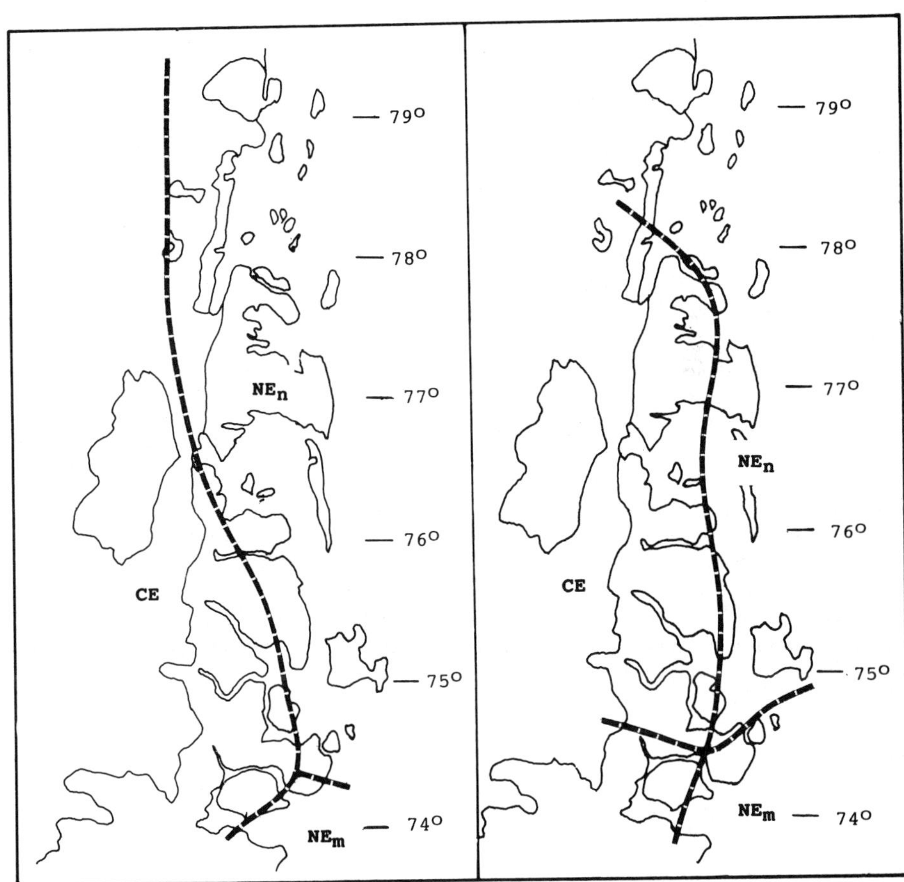

Fig. 24. Delimitation of the coastal and continental flora districts in Northeast Greenland according to Böcher *et al.* (1959) (left) and Seidenfaden & Sørensen (1937) (right).

(Map 81) (southern limit at 75°11′N) and *Polemonium boreale* (Map 218) (northern limit at 74°52′N). The northern limit of the middle district in Northeast Greenland may be found south of the present study area, possibly in the southern part of Hold with Hope where *Polemonium boreale* and *Saxifraga platysepala* (Map 62) have their known southern limits.

Elvebakk (1985) has subdivided the European Arctic and adjacent areas based on phytosociological criteria. His delimitation of the Middle (MATZ) and Northern Arctic Tundra Zone (NATZ) in Greenland is to be altered according to the present study. The border line between the MATZ and NATZ at the east coast proposed at Skærfjorden (app. 77°N) should be moved northwards to the north side of Lambert Land (app. 79°30′N) as many low and middle arctic species were found this far north in 1990.

6.2. Phytogeographical affinities

A comparison between the geographical distribution types (Table 6) of the total study area with data from the floristic province North Greenland, northern Ellesmere Island (Soper & Powell 1985, Hedderson 1990), Svalbard (Rønning 1971, Elvebakk pers. comm.), and the Russian High Arctic (Alexandrova 1988) (Table 10) shows a similarity in the frequency of circumpolar species (56–65%). The percentage of amphi-Atlantic and western species, however, is almost the opposite in Lake Hazen, Ellesmere Island, and North Greenland, the last mentioned having 24% of western species and 12% of amphi-Atlantic against 11% and 19%, respectively. Surprisingly, Lake Hazen has fewer western taxa than North Greenland. Borup Fjord, Ellesmere Island, has a nearly equal number of amphi-Atlantic and western species. As expected, Svalbard and Russia have the highest percentages of eastern species. Lake Hazen has a rather high proportion of endemic species, whereas North Greenland, Svalbard (Elvebakk 1985), and Russia have none. High Arctic Sovjet has the largest percentage of amphi-Beringian species.

In spite of the different delimitations of the districts and the fact that the present study area only includes the northern parts of the districts in Northwest and Northeast Greenland sensu Böcher *et al.* (1959), a comparison of the results of the present investigation and their results is possible. The frequency of western and eastern species is given in per centage of the total native flora of each district (Fig. 26). The frequencies of these floral elements in northern Greenland has only been slightly altered.

The highest per cent of western species is still in Northwest Greenland. The per cent of western species is lower than in 1959 although new western species have been found in the districts mainly in the last decade (*Ledum palustre, Pedicularis sudetica* ssp. *albolabiata, P. flammea, Puccinellia bruggemanni,* and *Sagina caespitosa* in Northwest Greenland, and *Saxifraga tricuspidata* in North Greenland). The decline in per cent may be ascribed to the general increase in number of species due to increased botanical activity. There is a marked decline in western species entering the Melville Bugt in the southern northwest district and a general decrease through North and Northeast Greenland. The coastal districts have lower per cent of western species compared to the inland districts. The influence of eastern species is greater in the northeast districts, especially in the southern part. The greatest difference between the present study and the study in 1959 is the lower percentage of eastern species in the districts at the east coast. Only between 2–8% of the species in a district at the east coast belong to the eastern flora element according to the present study contrasting 5–10% in the 1959 study.

Table 10. Frequency (%) of the geographical distribution in northern Greenland; the floristic province North Greenland; lake Hazen area, Ellesmere Island; Svalbard; and High Arctic Sovjet. Cir: circumpolar, Amp-A: amphi-Atlantic, W: western, E: eastern, Amp-B: amphi-Beringian, End: endemic.

	Cir	Amp-A	W	E	Amp-B	End
Northern Greenland (74°–83°40′N)	56	15	22	6	0,5	0,5
Northern Greenland (79°–83°40′N)	61	12	24	2	0,5	–
Lake Hazen[1] Canada (81°°45′–82°10′N)	60	19	11	–	2	8
Borup Fjord[2] Canada (81°50′N)	61	17	18	–	4	–
Svalbard[3] Norway (74°–81°N)	61	13	10	16	–	–
High Arctic Sovjet[4] (75°–81°N)	65	4	–	10	15	–

[1] From Soper & Powell 1985
[2] From Hedderson 1990
[3] From Rønning 1971, Elvebakk pers. comm.
[4] From Alexandrova 1988

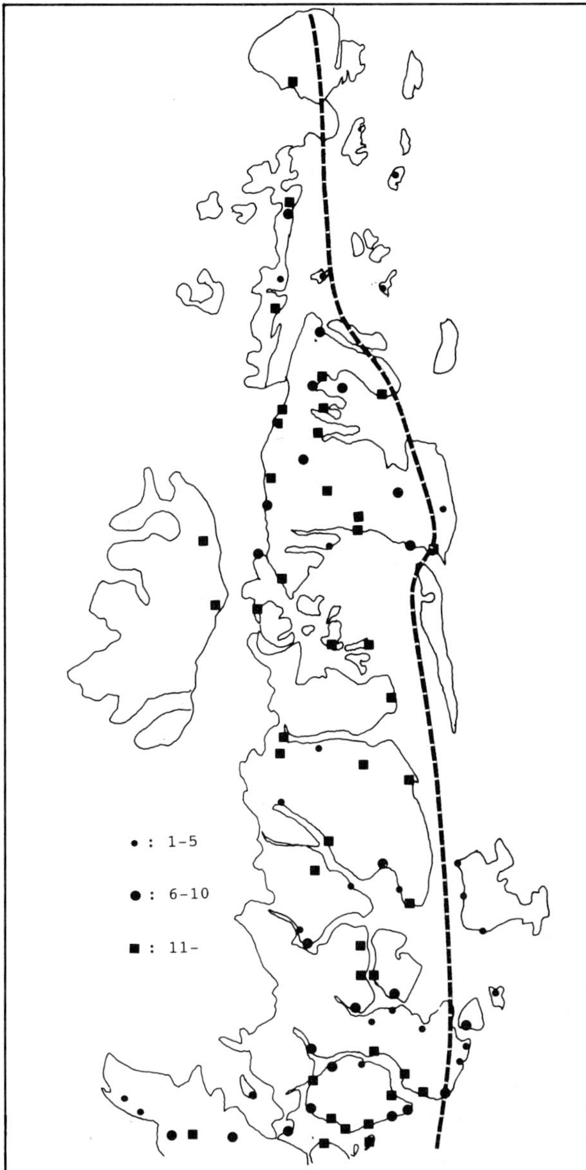

Fig. 25. Delimitation of coastal and continental areas in Northeast Greenland based on the present material. 39 species considered continental are mapped and number of species per locality is indicated. A border line between coastal and continental areas is proposed.

6.3. Discussion of the distribution patterns in northern Greenland

As competition is of minor importance to the distribution of plants in the Arctic (Savile 1960) the distribution of plants not occurring throughout the area must reflect physical conditions and/or historical events. The reduction in number of species with increasing latitude is primarily caused by changes in the climatic conditions. Phytogeographical limits in areas with no lithological borders must mainly be ascribed to climatic conditions or historical events. Thus, the marked floristic border at 79°N at both coasts is not reflecting a change in geology.

Generally, species only present in the southern part of the area are middle or low arctic, having a certain demand for summer warmth and length of growing season not fulfilled in the northern parts. These are mainly found in NGDT 5–14. Middle arctic species, which are absent in South Greenland, mainly reach 79°N, whereas low arctic species generally have a continuous occurrence from South Greenland to 72°30′N in West Greenland and to 75–76°N in East Greenland. The great difference in number of low arctic species reaching West Greenland northwards to 76°N (five spp.) and East Greenland (36 spp.) can be explained by a continental climate in the inland areas on the east coast contrary to the oceanic climate in the narrow coast land in Melville Bugt, only suitable for ubiquitous and coastal species. Many of the low arctic species only reaching East Greenland in the southernmost part of the study area are herb-slope species, which grow on south-facing slopes at the south coast of Clavering Ø, Wollaston Forland, and Kuhn Ø. On the contrary, the high arctic species not occurring south of 76°N on one or both of the coasts have a climatically conditioned border, not being able to grow in areas with a relatively high summer temperature i.e. higher than 3–4°C (NGDT 4).

Distribution patterns of species with a continuous distribution both in coastal and continental areas in some of the regions and to the south of the study area but missing in Melville Bugt can be ascribed to edaphic conditions (NGDT 2, 9 and parts of 8, 10). Gneissic bedrock dominates in Melville Bugt from Upernavik Isstrøm (72°30′N) to Kap Atholl (76°30′N) whereas alkaline sandstone occurs to the north and basalt is common to the south of Melville Bugt.

An isolated occurrence could represent either a recent immigration of the species, which has not yet spread to all potential habitats in the region, or a "wintering" during the last glaciation in un-glaciated areas from where it has not been able to spread.

Species with an otherwise wide distribution in the Arctic but with an isolated occurrence in Greenland in areas not glaciated during the Weichselian glaciation can be considered glacial relicts. Several areas in the northern part of Greenland have proved no or little evidence of glacial erosion: western part of Disko, Nuussuaq, and Svartenhuk (West Greenland), and southwestern part of Thule district, coastal areas in Washington Land, Hall Land, Nyeboe Land, Peary Land (North Greenland), and Jameson Land, Hold with Hope, Hochstetter Forland, and Shannon (East Greenland) (summarized by Funder 1989). Species with a present distribution that roughly coincide with four or three of these areas are: *Carex atrofusca* (Map 156), *Ranunculus nivalis* (Map 147), *Puccinellia andersonii* (Map 158), *Elymus hyperarcticus* (Map 135), *Carex marina* ssp. *pseudolagopina* (Map 157), and *Minuartia stricta* (Map

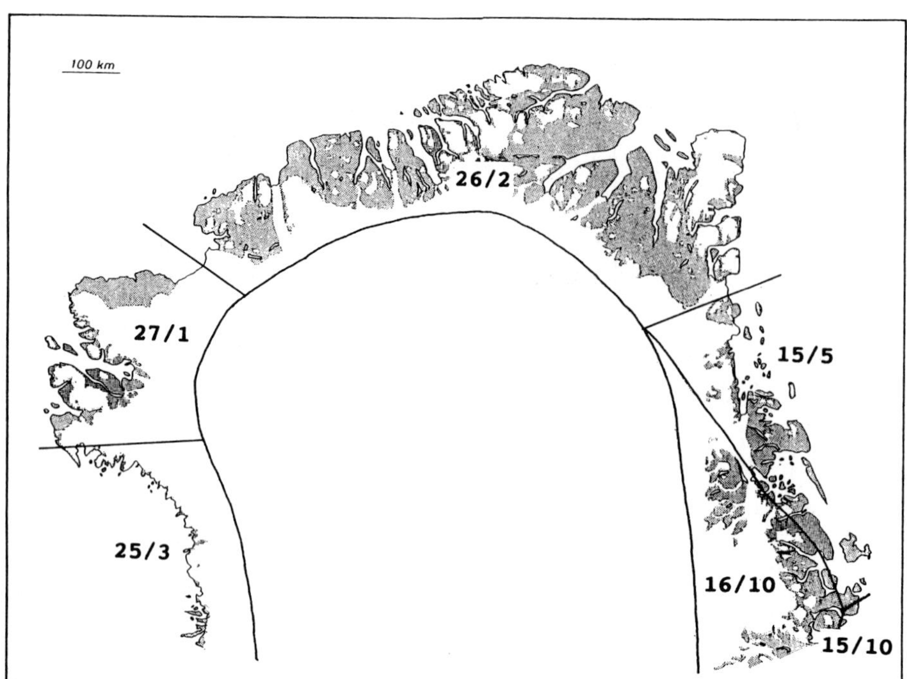

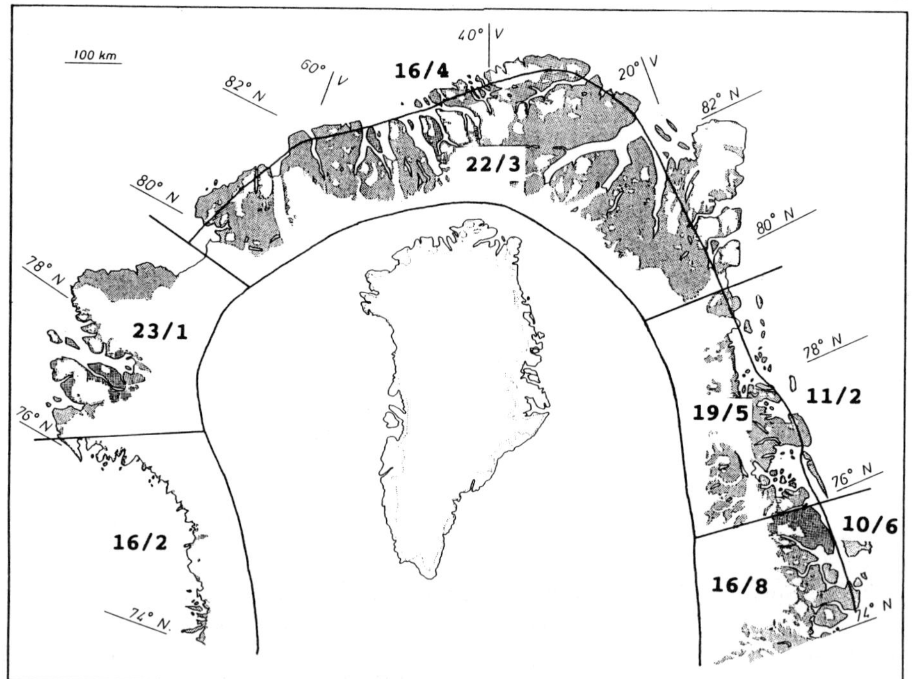

Fig. 26. A comparison of the western and eastern flora element in Böcher *et al.* (1959) (upper) and the present study (lower). The frequencies of western and eastern species are indicated in percentage of the total native flora of each district. The first figure indicates western, second figure eastern species.

142) of which the two last mentioned also have isolated occurrences in the Søndre Strømfjord area. These species may be considered glacial relicts.

The knowledge of the distribution of many of the unicentric and bicentric species in East Greenland used by Gelting (1934) in arguing for the existence of ice-free refugia has only been slightly altered as a result of the additional floristic information gained in the last decade. Unicentric species as *Polemonium boreale* (Map 218), *Potentilla stipularis* (Map 215), *Matricaria maritima* ssp. *phaeocephala* (Map 212), *Luzula wahlenbergii* (Map 211), *Chrysosplenium tetrandrum* (Map 210), *Carex vaginata* (Map 209), *Carex parallela* (Map 208), and the bicentric species *Dupontia psilosantha* (Map 206), *Braya linearis* (Map 205), and *Arctostaphylos alpina* (Map 204) occur in restricted areas close to or in the

Fig. 27. High arctic distribution types in Greenland.

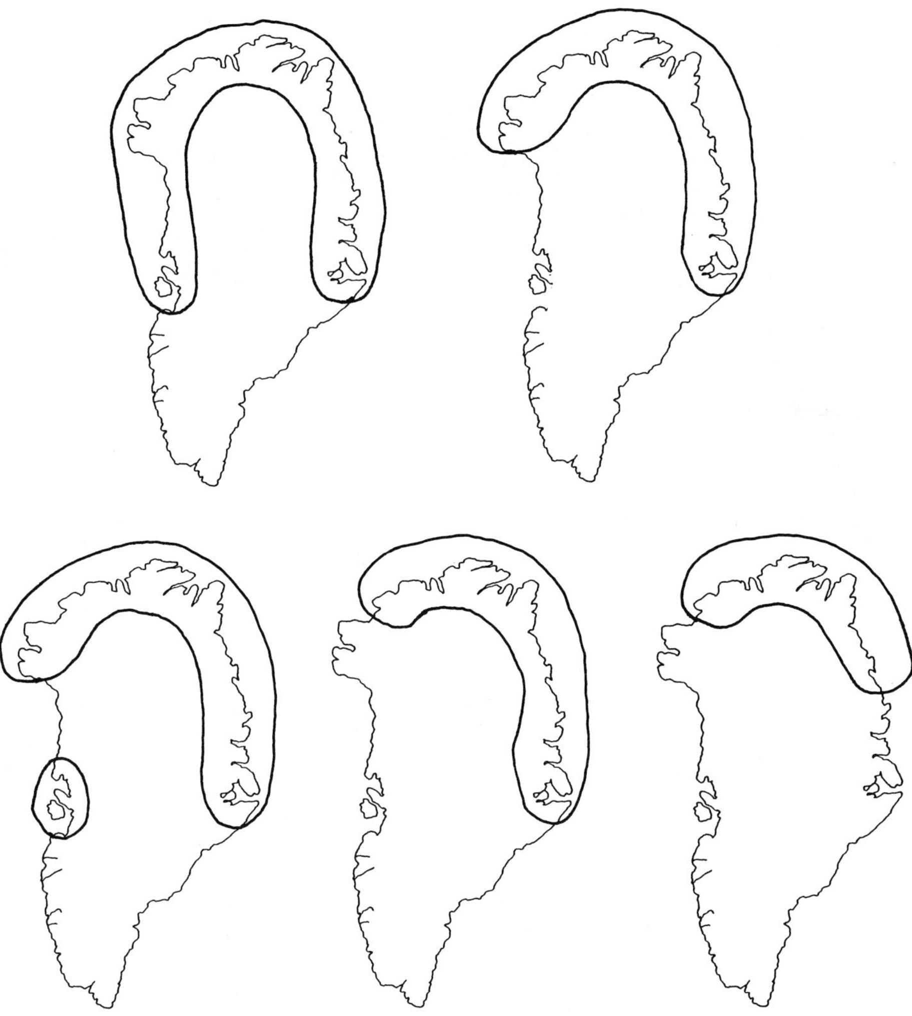

areas supposed to have been unglaciated during the last glaciation: Jameson Land, Hold with Hope, and Hochstetter Forland. This distribution pattern for at least the unicentric taxa, can still be explained by their surviving in these areas during glaciation combined with a weak dispersal ability.

A more likely explanation for the presence of *Chrysosplenium tetrandrum* and *Matricaria maritima* ssp. *phaeocephala* in few isolated sites in Northeast Greenland, mostly associated with the former Eskimo habitation, is that they have been introduced by Eskimos from Arctic North America on their migration around the north of Greenland (Sørensen 1945), a parallel to the finding of *Elymus mollis* at an Eskimo site 1000 km to the north of its known distribution area (Porsild & Cody 1979).

Some of the species only known from Thule district (NGDT 10) probably immigrated from the Canadian High Arctic islands in a fairly recent time. Species as *Androsace septentrionalis* (Map 167), *Pedicularis langsdorffii* (Map 169), and *Pedicularis sudetica* ssp. *albolabiata* (Map 170) are only known from very few localities within the former glaciated areas in Northwest Greenland and are common throughout most parts of arctic North America.

The distribution of the extremely high arctic species, which occur only in Thule district and North Greenland (NGDT 4), may be considered as caused by climatic conditions. On the other hand this type could either represent a relict occurrence of species, which has not been able to expand since the glaciation, or species of recent immigration. The distribution of the bicentric *Braya thorild-wulffii* (Map 82) confirmed by the isolated occurrences on Nuussuaq and Disko may represent a glacial relict whereas other species of this distribution type: *Taraxacum hyparcticum* (Map 87), *Pedicularis capitata* (Map 85), and *Erysimum pallasii* (Map 84), might be more recent immigrants. Their distribution outside Greenland with occurrences far to the south of their southern limit in Greenland implies that these species are not restricted to areas with an extremely cold summer climate.

In North Greenland five distribution types considered high arctic judging from their distribution in Greenland are recognized: 1) species widespread in northern Greenland, having southern limits at app. 68–70°N, 2) species widespread but missing in Melville Bugt. Their southern limit in West Greenland is in the Disko-Nuussuaq area (69–70°N) and Scoresby Sund (70°N) in East Greenland, 3) species widespread north of Melville Bugt (76°N), through Northwest and North Greenland to Scoresby Sund, 4) species only occurring in North and Northeast Greenland, north of Humboldt Gletscher (79°N), and 5) species only occurring in North Greenland north of Melville Bugt and Dove Bugt (76°N) (Fig. 27).

7. Acknowledgements

The present study was rendered possible by a two and a half year grant from the University of Copenhagen. The field work in 1986–88 and 1988–90 was supported by the Commission for Scientific Research in Greenland and Aage V. Jensen Charity Foundation, respectively. The Danish Natural Science Research Council sponsored the field work in 1991. Greenland Geological Survey, the Danish Air Force, the sledge patrol Sirius, Danish Polar Center, and the Danish Meteorological Institute provided invaluable logistic support in the field, and in preparing the field work in 1985–87 and 1989–91. The support of these institutions are gratefully acknowledged.

Several persons have contributed with valuable plant collections, among others Ole Bennike and Svend Funder, Geological Museum, Copenhagen; Per Mølgaard, Royal Danish School of Pharmacy; Geoffrey Halliday, University of Lancaster; Lars Stemmerik, Greenland Geological Survey; T. Hauge Andersson, Danish Polar Center; Roger Goodwillie; R. Corner; Sue and Rob David, and H. Lang. All are thanked for the collecting work and for placing their material at my disposal.

Finally, I am deeply indebted to my supervisor Bent Fredskild, Botanical Museum, Copenhagen, and to Eilif Dahl, Oslo and Kjeld Hansen, København for their critical reading of the manuscript and valuable comments, to Birthe Hammer for correcting the English and to Lisbet Ellekilde for the design of figures and technical assistance.

Addendum

Gentiana detonsa Rottb., hitherto not known north of Ella Ø (72°53′N) in East Greenland, was found at Tyroler Fjord app. 74°30′N during the British North East Greenland Project Clavering Island Expedition 1992. The specimen which has been verified by me was placed at my disposal by the expedition leader Rob David.

8. References

Aastrup, P., Bay, C. & Christensen, B. 1986. Biologiske miljøundersøgelser i Nordgrønland 1984–85. – Grønlands Fiskeri- og Miljøundersøgelser: 11–58.

Alexandrova, V. D. 1980. The Arctic and Antarctic: Their division into Geobotanical Areas. – Cambridge University Press: 247 pp.

– 1988. Vegetation of the Soviet polar deserts. – Cambridge University Press: 228 pp.

Babb, T. A. & Bliss, L. C. 1974. Effects of physical disturbance on arctic vegetation in the Queen Elisabeth Islands. – J. Appl. Ecol. 11: 549–562.

Bay, C. 1983. En plantegeografisk undersøgelse af Nordvestgrønland. Thesis, University of Copenhagen: 238 pp.

– 1991. Botanical studies. – In: Andreasen, C., Angantyr, L. A., Bay, C., Boertmann, D., Born, E. W., Elling, H., Helms, H. J., Larsen, F., Olesen, C. R. & Siegstad, H. Nature conservation in Greenland: 66–77. – Atuakkiorfik, Copenhagen.

– in prep. Taxa of vascular plants new to the flora of Greenland.

– & Holt, S. 1986. Vegetationskortlægning af Jameson Land 1982–86: 40 pp. – Grønlands Fiskeri- og Miljøundersøgelser.

– & Boertmann, D. 1989. Biologisk-arkæologisk kortlægning af Grønlands østkyst mellem 75°N og 79°30′N. Del 1: Flyrekognoscering mellem Mesters Vig (72°12′N) og Nordmarken (78°N). – Grønlands Hjemmestyre, Miljø- og Naturforvaltning, Teknisk rapport 4: 63 pp.

– & Fredskild, B. 1990. Biologisk-arkæologisk kortlægning af Grønlands østkyst mellem 75°N og 79°30′N. Del 3: Botaniske undersøgelser i området mellem Fligely Fjord (74°50′N) og Nordmarken (77°30′N), 1989. – Grønlands Hjemmestyre, Miljø- og Naturforvaltning, Teknisk Rapport 11: 56 pp.

– & Fredskild, B. 1991. Biologisk-arkæologisk kortlægning af Grønlands østkyst mellem 75°N og 79°30′N. Del 6: Botaniske undersøgelser i området mellem Germania Land (77°N) og Lambert Land (79°10′N), 1990. – Grønlands Hjemmestyre, Miljø- og Naturforvaltning, Teknisk Rapport 24: 55 pp.

Bennike, O. 1990. The Kap København Formation: stratigraphy and palaeobotany of a Plio-Pleistocene sequence in Peary Land, North Greenland. – Meddr Grønland, Geosci. 23: 85 pp.

Bessels, E. 1879. Die Amerikanische Nordpol-Expedition. – Leipez: 647 pp.

Blake, W. Jr., Bouclerke, M., Fredskild, B., Janssens, J. A. & Smol, J. P. 1992. The Holocene history of "Kap Inglefield Sø", Inglefield Land, northwestern Greenland. Meddr Grønland, Geosci. 27: 42 pp.

Bliss, L. C. & Svoboda, J. 1984. Plant communities and plant production in the western Queen Elizabeth Islands. – Holarct. Ecol. 7: 325–344.

Bonneval, L. de 1977. Quelques indications sur la vegetation de Cass Fj., Washington Land. – Unpublished report: 16 pp.

Buchenau, F. & Focke, W. O. 1874. Gefässpflanzen. In: Die zweite Deutsche Nordpolarfahrt in den Jahren 1869–70. Vol. 2. Wissenschaftlische Ergebnisse. Leipzig.

Böcher, T. W. 1951. Studies on the distribution of the units within the collective species of *Stellaria longipes*. – Bot. Tidsskr. 48 (4): 401–420.

– 1966. Experimental and Cytological Studies on plant species. IX. Some arctic and montane Crucifers. – Biol. Skr. Vid. Selsk. 14 (7): 74 pp.

– 1975. Det grønne Grønland. – København, 256 pp.

– , Holmen K. & Jakobsen, K. 1959. A synoptical study of the Greenland flora. – Meddr Grønland 163 (1): 32 pp.

– , Fredskild, B., Holmen K. & Jakobsen, K. 1966. The flora of Greenland (2nd edition). – København, 312 pp.

– , Fredskild, B., Holmen, K. & Jakobsen, K. 1978. Grønlands Flora (3rd edition). – København, 327 pp.

Cabot, D., Goodwillie, R. & Viney, M. 1988. Irish Expedition to Northeast Greenland 1987. – Dublin 37–73.
Eastwood, A. 1948. Botanical collections. – In: Boyd, L. (ed.). The coast of Northeast Greenland: 270–273. – New York.
Edlund, S. A. 1983. Bioclimatic zonation in a High Arctic region: Central Queen Elisabeth Islands. – In: Current Research, part a, Geological Survey of Canada, paper 83–1A: 381–390.
Ekblaw, W. E. 1919. The plant life of northwest Greenland. – Natural history 19: 272–291.
Ekman, E. 1933. Contribution to the flora of Greenland V. – Sv. Bot. Tidsskr. 27, 1: 102–103.
Elkington, T. T. 1965. Studies on the variation of the genus *Dryas* in Greenland. – Meddr Grønland 178 (1): 56 pp.
Elvebakk, A. 1985. Higher phytosociological syntaxa on Svalbard and their use in subdivision of the Arctic. – Nord. J. Bot. 5: 277–284.
Feilberg, J. 1984. A phytogeographical study of South Greenland. Vascular plants. – Meddr Grønland, Biosci. 15: 70 pp.
Fredskild, B. 1966a. Contributions to the flora of Peary Land, North Greenland. – Meddr Grønland 178 (2): 22 pp.
– 1966b. Den arktiske ørken. – Naturens Verden 1966: 1–16.
– 1973. Studies in the vegetational history of Greenland. – Meddr Grønland 198 (4): 247 pp.
– 1985. The Holocene vegetational development of Tugtuligssuaq and Qeqertat, Northwest Greenland. – Meddr Grønland, Geosci. 14: 20 pp.
– in prep. A phytogeographical study of West Greenland between 62°20′N and 74°N.
– , Feilberg, J., Hanfgarn, S. & Alstrup, V. 1979. Grønlands Botaniske Undersøgelse 1978 og 1979: 27–31. – Botanisk Museum, København.
– & Bay, C. 1980. Botanical investigations. – In: Bay, C., Fredskild, B., Grønnow, B., Jakobsen, B. H., Meldgaard, M., Mortensen, N. G. & Thingvad, N. Report of activities at Tugtuligssuaq, Melville Bugt, 1978–79. – Geogr. Tidsskr. 80: 39–41.
– & Møller, M. 1981. Grønlands Botaniske Undersøgelse 1980–81: 10–13. – Botanisk Museum, København.
– , Bay, C., Holt, S. & Nielsen, B. 1986. Grønlands Botaniske Undersøgelse 1985: 26–33. – Botanisk Museum, København.
– , Bay, C. & Petersen, F. R. 1987. Grønlands Botaniske Undersøgelse 1986: 27–33. – Botanisk Museum, København.
– & Bay, C. 1988. Grønlands Botaniske Undersøgelse 1987: 27–33. – Botanisk Museum, København.
– & Bay, C. 1989. Grønlands Botaniske Undersøgelse 1988: 18–26. – Botanisk Museum, København.
– & Mogensen, G. S. 1991. Botanik: Vegetationen omkring Zackenberg. – In: Andersson, T. I. H., Böcher, J., Fredskild, B., Jakobsen, B. H., Meltofte, H., Mogensen, G. S. & Muus, B. 1991. Rapport om muligheden for placering af en naturvidenskabelig forskningsstation ved Zackenberg, Nationalparken i Nord- og Østgrønland: 22–35. – Dansk Polarcenter, København.
– , Bay, C., Feilberg, J. & Mogensen, G. S. 1992. Grønlands Botaniske Undersøgelse 1991: 24–34. – Botanisk Museum, København.
Fristrup, B. 1952. Physical geography of Peary Land. 1. Meteorological observations for Jørgen Brønlund Fjord. – Meddr Grønland 127 (4): 143 pp.
Fränkl, E. 1955. Rapport über die Durchquerung von Nord Peary Land (Nordgrönland) im Sommer 1953. – Meddr Grønland 103 (8): 31–32.
Funder, S. 1989. Quartenary Geology of the ice-free and adjacent shelves of Greenland. – In: Fulton, R. J. (ed.) Quartenary Geology of Canada and Greenland. – Geological Survey of Canada: 741–792.
– & Abrahamsen, N. 1988. Palynology in a polar desert, eastern North Greenland. – Boreas 17: 195–207.
Gelting, P. 1934. Studies on the vascular plants of East Greenland between Franz Joseph Fjord and Dove Bay (Lat. 73°15′–76°20′N). – Meddr Grønland 101 (2): 337 pp.
– 1937. Studies of the food of the East Greenland ptarmigan especially in its relation to vegetation and snow-cover. – Meddr Grønland 116 (3): 196 pp.
Ghisler, M. 1986. Geology and mineral deposits in the National Park in North-east Greenland. – Forskning/tusaat 3/84: 43–51.
Greely, A. 1886. Three years of arctic service, on account of the lady Franklin Bay Expedition 1881–84. – London.
Griffin, A. 1972. Botany Report. – In: Peacock, J. D. C. (ed.) Joint Services Expedition to North Peary Land 1969: Annex C.
Halliday, G. 1981. British North-east Greenland Expedition 1980: 20–28. – University of Lancaster.
– in prep. The nunatak flora of Southeast Greenland.
– & Corner, R. W. M. 1990. Botanical investigations on Th. Thomsen Land and on Kuhn Ø. – In: Fredskild, B. & Bay, C. 1991. – Grønlands Botaniske Undersøgelse 1990: 34–36. – Botanisk Museum, København.
Hansen, B. U. & Søgaard, H. 1989. Vegetations- og sne/iskortlægning af den grønlandske Nationalpark 1988: 28 pp. – Geografisk Institut, Københavns Universitet.
Hansen, E. S. 1987. Lichenerne i Qaanaaqs vegetationsmosaik. – Forskning/tusaat 2/87: 28–39.
– 1989. The lichen flora of Qaanaaq (Thule), Northwest Greenland. – Mycotaxon 35 (2): 379–394.
Hart, H. 1880. On the botany of the British Polar Expedition of 1875–6. – Journal of Botany, N. 9: 52–56, 70–79, 111–115, 141–145, 177–182, 204–208, 235–242, 303–306.
Hartz, N. & Kruuse, C. 1911. The vegetation of Northeast Greenland. 69°25 lat. n. –75° lat. n. – Meddr Grønland 30 (10): 333–431.
Hedderson, T. A. 1990. Floristics and Phytogeography. – In Hankinson, K. W. (ed.) 1990: Joint Services Expedition to Borup Fjord, Ellesmere Island 1988, Part 2.
Heide, O. M., Pedersen, K. & Dahl, E. 1990. Environmental control of flowering and morphology in the high-arctic *Cerastium regelii,* and the taxonomic status of *C. jenisejense.* – Nord. J. Bot. 10: 141–147.
Henry, G., Fredman, B. & Svoboda, J. 1986. Survey of vegetated areas and muskox populations in East-Central Ellesmere Island. – Arctic 39 (1): 78–81.
Holmen, K. 1955. Notes on the bryophyte vegetation of Peary Land, North Greenland. – Mitt. Thyr. Bot. Gesell. 1 (2–3): 96–106.
– 1957. The vascular plants of Peary Land, North Greenland. – Meddr Grønland 124 (9): 149 pp.
Hooker, W. J. 1825. Some account of a collection of arctic plants formed by Edvard Sabine. – Transact. Linnean. Soc., 14.
Hultén, E. 1956. The *Cerastium alpinum* complex. A case of world-wide introgressive hybridization – Sv. Bot. Tidsskr. 50 (3): 411–495.
– 1958. The amphi-Atlantic plants and their phytogeographical connections. – K. Svenska Vetensk. – Akad. Handl. 4,7,1: 340 pp.
– 1962. The circumpolar plants. I. – K. Svenska Vetensk. – Akad. Handl. 4,7,5: 280 pp.
– 1968. Flora of Alaska and neighboring territories. – Stanford University Press, California: 1008 pp.
– 1971. The circumpolar plants. II. – K. Svenska Vetensk. – Akad. Handl. 4,13,1: 463 pp.
– & Fries, M. 1986. Atlas of North European vascular plants, I-III. – Koeltz Sci. Books, Königstein: 1172 pp.
Jakobsen, B. H. 1988. Soil formation on the Peninsula Tugtugligssuaq, Melville Bugt, North West Greenland. – Geogr. Tidsskr. 88: 86–93.
– , Mortensen, N. G. & Thingvad, N. 1980. – In: Bay, C., Fredskild, B., Grønnow, B., Jakobsen, B. H., Meldgaard, M., Mortensen, N. G. & Thingvad, N. Report of activities at Tugtuligssuaq, Melville Bugt, 1978–79. – Geogr. Tidsskr. 80: 29–44.

Jakobsen, K. 1950. Unpublished notes from the field work in Northwest Greenland, 1950.
Klein, D. R. & Bay, C. 1990. Fouraging dynamics of muskoxen in Peary Land, northern Greenland. – Holarct. Ecol. 13: 269–280.
– & Bay, C. in prep. Herbivores in the High Arctic: Adaptations to constraints on plant growth.
Kruuse, C. 1905. List of the phanerogams and vascular cryptogams found on the coast 75°–66°29′ lat. N of East Greenland. – Meddr Grønland 30 (4): 143–209.
Lange, J. 1887. Bemærkninger om de af Ekspeditionerne i Aarene 1886–87 samlede Karplanter fra Vestkysten af Grønland. – Meddr Grønland 8 (9): 281–288.
Lundager, A. 1912. Some notes concerning the vegetation of Germania Land, North-east Greenland. – Meddr Grønland 43, (13): 349–414.
Löve, A. & Löve, D. 1961. Some nomenclatural changes in the European Flora. – Bot. Notiser 114: 48–56.
Mogensen, G. S. 1985. Vegetationen ved Kap København. – Naturens Verden 1985, 6–7: 241–244.
Mosquin, T. & Hayley, D. E. 1966. Chromosome numbers and taxonomy of some Canadian Arctic plants. – Can. J. Bot. 44: 1214.
Mortensen, N. G. 1987. Unpublished meteorological data from Kap Moltke, Nordgrønland.
Muc, M., Freedman, B., & Svoboda, J. 1989. Vascular plant communities of a polar oasis at Alexandra Fiord (79°N), Ellesmere Island, Canada. – Can. J. Bot. 67: 1126–1136.
Mølgaard, P. 1985. The Kilen Expedition 1985 – a first account of the geological and biological results. – Unpublished report.
Oosting, H. J. 1948. Ecological notes on the flora. – In: Boyd, L. (ed.). The coast of Northeast Greenland: 225–270. – New York.
Ostenfeld, C. H. 1905. A list of flowering plants from Cape York and Melville Bugt (N. W. Greenland) collected by the Rev. Knud Balle and Mr. L. Mylius-Erichsen in 1903–05. – Meddr Grønland 33: 61–68.
– 1915. Plants collected during the First Thule Expedition to northernmost Greenland. – Meddr Grønland 51 (10): 371–381.
– 1919. Plante- og dyreliv på Grønlands nordkyst. – In: Rasmussen, K. Grønland langs Polhavet: 531–552.
– 1923a. Critical notes on the taxonomy and nomenclature of the flowering plants from northern Greenland. – Meddr Grønland 64: 163–188.
– 1923b. The vegetation on the north-coast of Greenland based upon the late Dr. Th. Wulff's collections and observations. – Meddr Grønland 64: 222–268.
– 1923c. Flowering plants and ferns from Wolstenholme Fjord (ca. 76°30′N lat.). – Meddr Grønland 64: 191–206.
– 1923d. Two plant lists from Inglefield Gulf and Inglefield Land. – Meddr Grønland 64: 209–214.
– 1925a. Vegetation of North Greenland. – Bot. Gazette 8 (2): 213–218.
– 1925b. Om plantevæksten på Grønlands nordkyst og dens livskår. – Naturens Verden 1925: 289–312.
– 1925c. Flowering plants and ferns from North-Western Greenland collected during the Jubilee Expedition 1920–22 and some remarks on the phytogeography of North-Greenland. – Meddr Grønland 68: 42 pp.
– & Lundager, A. 1910. List of vascular plants of North-East Greenland (N of 76° N lat.). – Meddr Grønland 43 (1): 32 pp.
Philipp, M. 1972. The *Stellaria longipes* group in N. W. Greenland. Cytological and morphological investigations. – Bot. Tidsskr. 67 (1–2): 64–75.
Porsild, A. E. 1955. The vascular plants of the western Canadian Arctic Archipelago. – Nat. Mus. Canada, Bull. 135: 226 pp.
– 1964. Illustrated flora of the Canadian Arctic Archipelago. – Nat. Mus. Canada, Bull. 146: 218 pp.
– 1965. The genus *Antennaria* in eastern Arctic and subarctic America. – Bot. Tidsskr. 61: 22–55.
– & Cody, W. J. 1979. Vascular plants of continental Northwest Territories, Canada. National Museums of Canada, Ottawa: 667 pp.
Powell, J. M. 1971. A list of vascular plants from Polaris Bay, Northwest Greenland. – Arctic 24 (2): 139–141.
Rønning, O. 1971. Synopsis of the flora of Svalbard. – Årb. norsk Polarinst. 1969: 80–93.
Savile, D. B. O. 1960. Limitations of the competitive exclusion principle. – Science 132: 1761 pp. – Washington.
Schwarzenbach, F. H. 1961. Botanische Beobachtungen in der Nunatakkenzone Ostgrönlands zwischen 74° und 75°N Br. – Meddr Grønland 163 (5): 172 pp.
Scoggan, H. J. 1978–79. The Flora of Canada. – Nat. Mus. Can. Pubs. Natur. Sci. Botany 7 (1–4): 1711 pp.
Seidenfaden, G. 1930. Botanical investigations during the Danish East Greenland Expedition 1929. – Meddr Grønland 74 (15): 366–382.
– 1931. Moving soil and vegetation in East Greenland. – Meddr Grønland 87 (2): 21 pp.
– 1932. The Godthåb Expedition 1928. – Meddr Grønland 82 (1): 16 pp.
– & Sørensen, T. 1933. On *Eriophorum callitrix* Cham. in Greenland. – Meddr Grønland 101 (1): 27 pp.
– & Sørensen, T. 1937. The vascular plants of Northeast Greenland from 74°30′ to 79°00′N lat. and a summary of all species found in East Greenland. – Meddr Grønland 101 (4): 215 pp.
Simmons, H. G. 1904: Notes of some rare or dubious Danish Greenland plants. – Meddr Grønland 26: 469–473.
– 1909. A revised list of the flowering plants and ferns of North West Greenland. – Report of The Second Norwegian Expedition in the "Fram" 1898–1902. No. 6: 110 pp.
Skvortsov, A. K. 1971. Zur Taxonomie der Weiden von Grönland, Island und den Fäeröerne. – Ann. Naturhistor. Mus. Wien. 75: 223–233.
Smith, H. 1914. *Catabrosa concinna* Th. Fr. *algidiformis* Nov. subsp. und ihre Nächstverwandten. – Sv. Bot. Tidsskr. 8: 245–252.
Soper, J. D. 1940. Local distribution of eastern Canadian birds. – Auk 57: 13–21.
Soper, J. H. & Powell, J. M. 1985. Botanical studies in the lake Hazen region, northern Ellesmere Island, Nortwest Territories, Canada. – Nat. Mus. Can. Pubs. Natur. Sci. No. 5: 67 pp.
Svoboda, J. & Freedman, B. 1981. Ecology of High Arctic lowland oases: Alexandra Fjord (78°53′N, 75°55′W), Ellesmere Island, N. W. T., Canada. – Monograph. Department of Botany, University of Toronto: 245 pp.
Sørensen, T. 1933. The vascular plants of East Greenland from 71°00′ to 73°30′N lat. – Meddr Grønland 101 (3): 177 pp.
– 1935. Bodenformen und Plantzendecke in Nordostgrönland. – Meddr Grønland 93 (4): 69 pp.
– 1941. Temperature relations and phenology of the Northeast Greenland flowering plants. – Meddr Grønland 125 (9): 305 pp.
– 1943. The flora of Melville Bugt. – Meddr Grønland 124 (5): 70 pp.
– 1945. Summary of the botanical investigations in N. E. Greenland. – Meddr Grønland 144 (3): 48 pp.
– 1954. New species of *Hierochloë, Calamagrostis,* and *Braya.* – Meddr Grønland 136 (8): 24 pp.
Tedrow, J. C. F. 1966. Polar desert soils. – Soils science society of America proceedings 30 (3): 381–387.
– 1970. Soil investigation in Inglefield Land, Greenland. – Meddr Grønland 188 (3): 13 pp.
Tsvelev, N. N. 1983. Grasses of the Soviet Union, part II: 735–737. – Leningrad.
Vaage, J. 1932. Vascular Plants from Eirik Raude's Land (East Greenland 71°30–75°40 lat. N). – Skrifter om Svalbard og Island 48: 87 pp.
Young, S. B. 1971. The vascular flora of the St. Lawrence Island with special reference to floristic zonation in the Arctic regions. – Contributions from the Gray Herbarium 201: 11–115.

Appendix I.

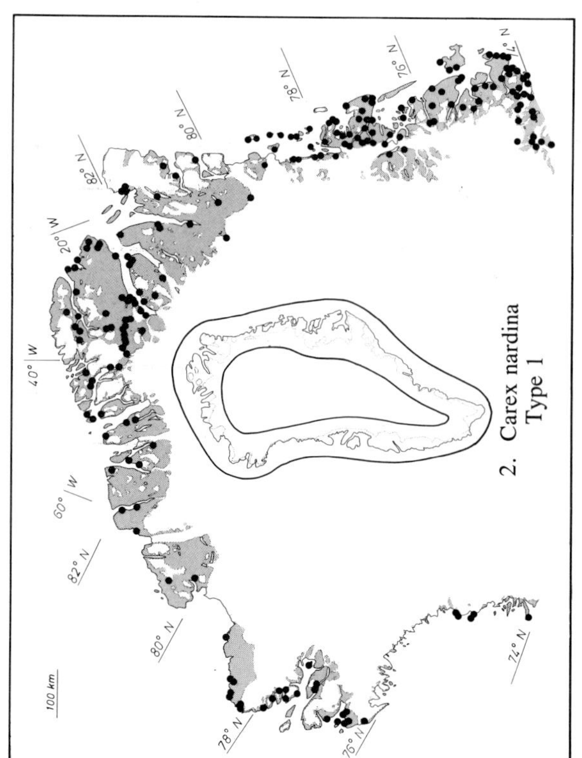

2. Carex nardina
Type 1

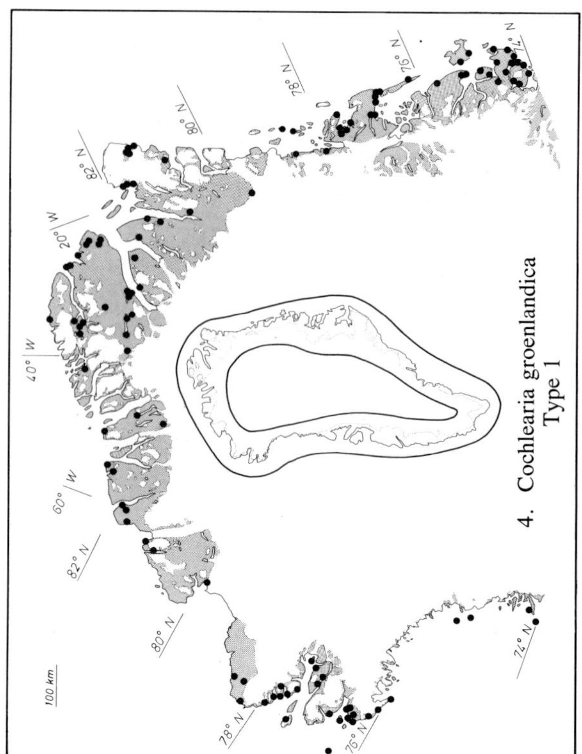

4. Cochlearia groenlandica
Type 1

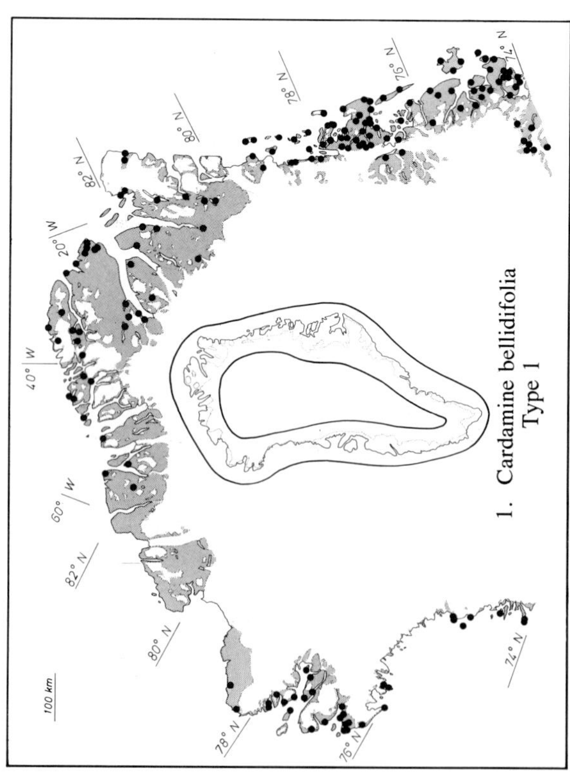

1. Cardamine bellidifolia
Type 1

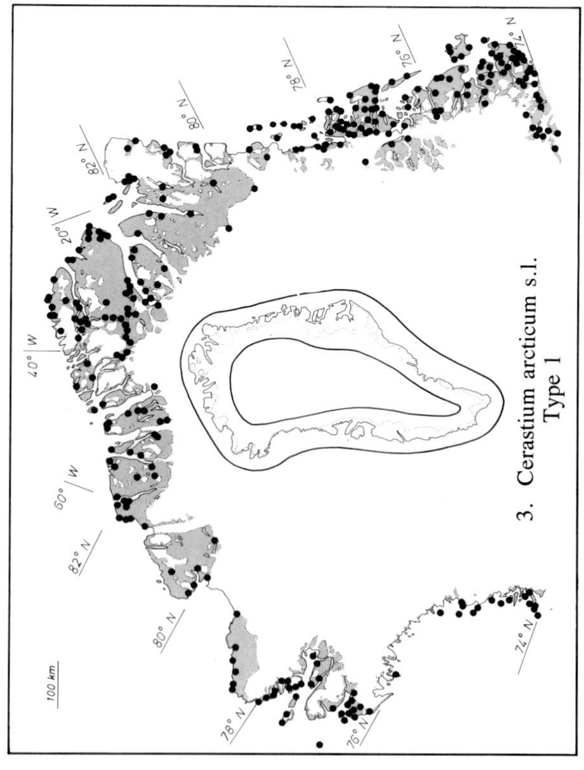

3. Cerastium arcticum s.l.
Type 1

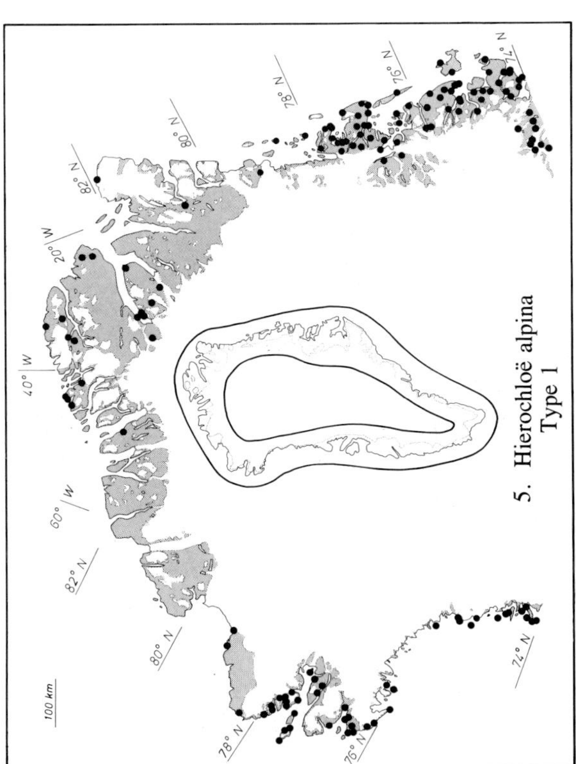

5. Hierochloë alpina
Type 1

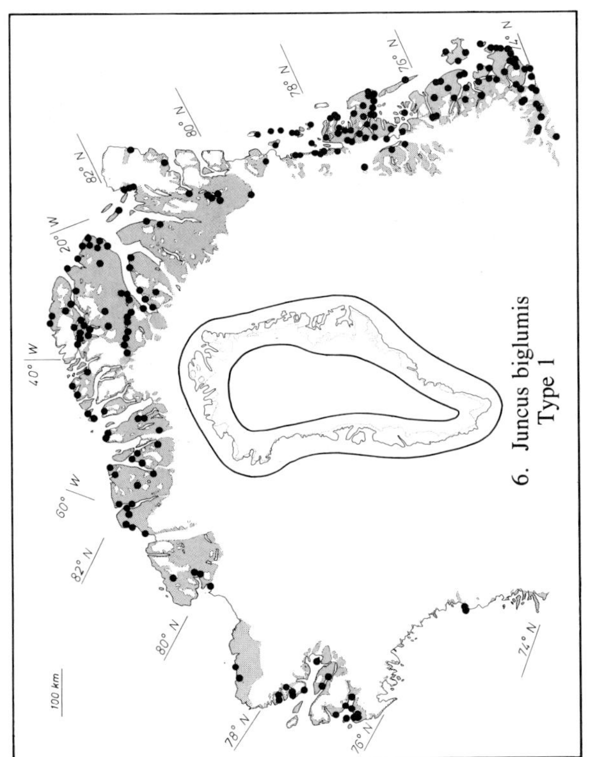

6. Juncus biglumis
Type 1

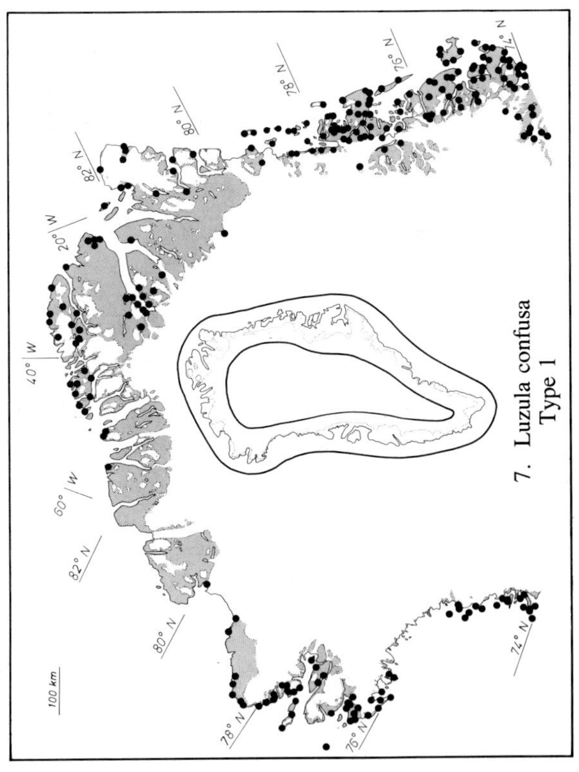

7. Luzula confusa
Type 1

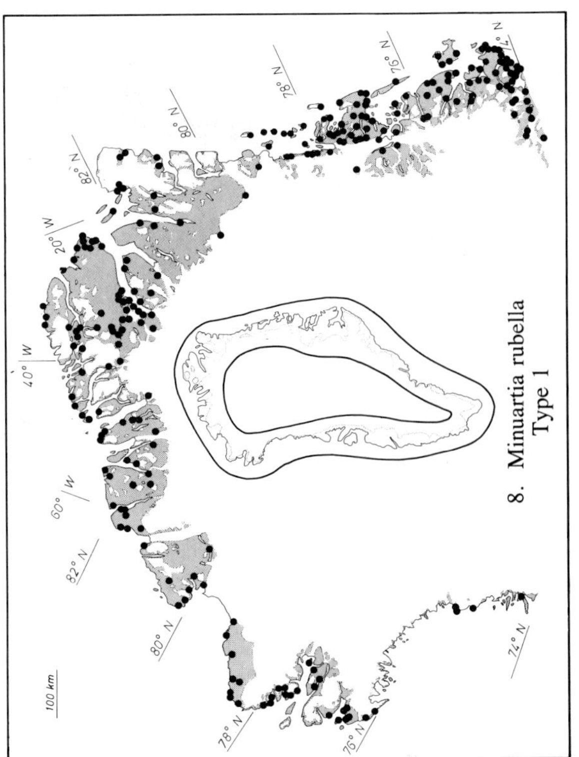

8. Minuartia rubella
Type 1

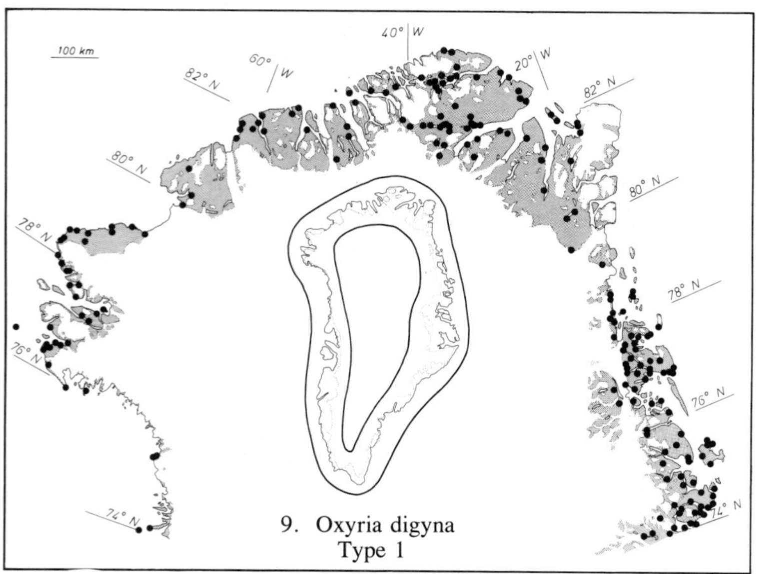

9. Oxyria digyna
Type 1

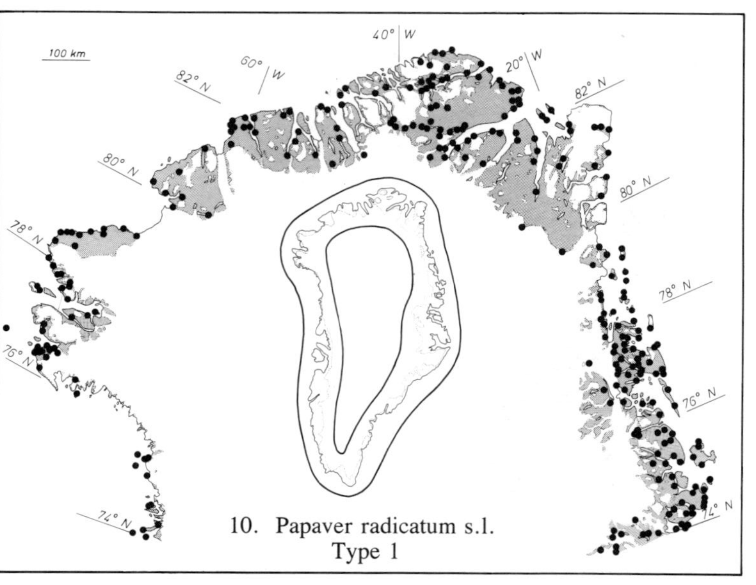

10. Papaver radicatum s.l.
Type 1

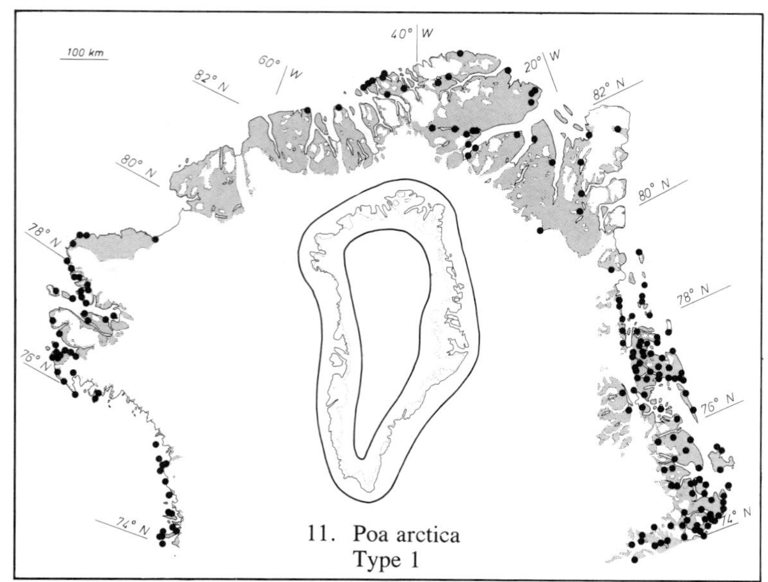

11. Poa arctica
Type 1

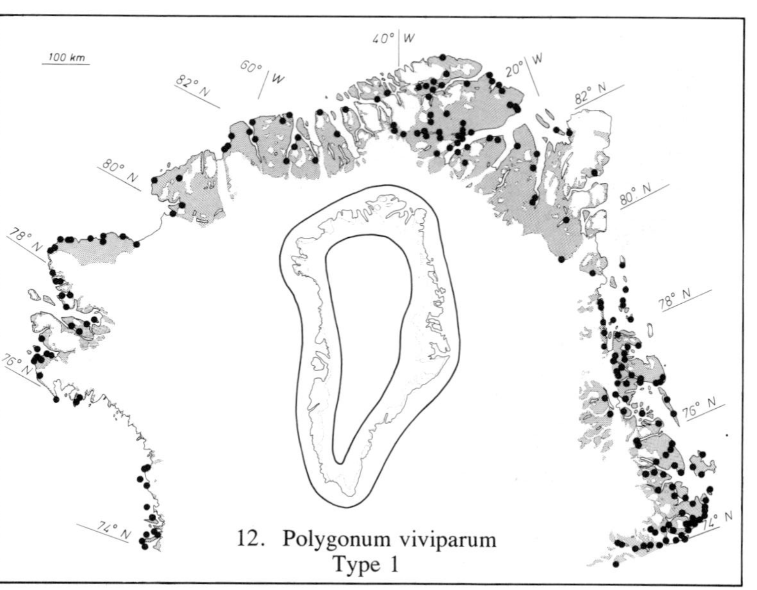

12. Polygonum viviparum
Type 1

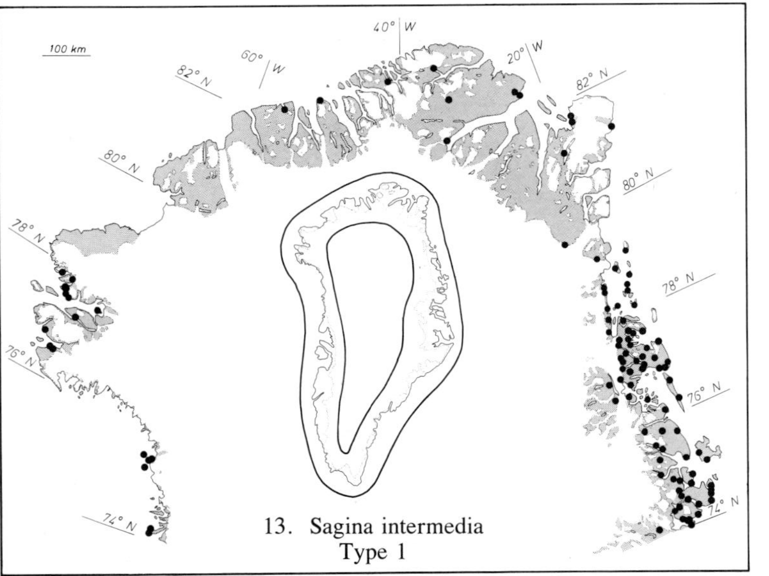

13. Sagina intermedia
Type 1

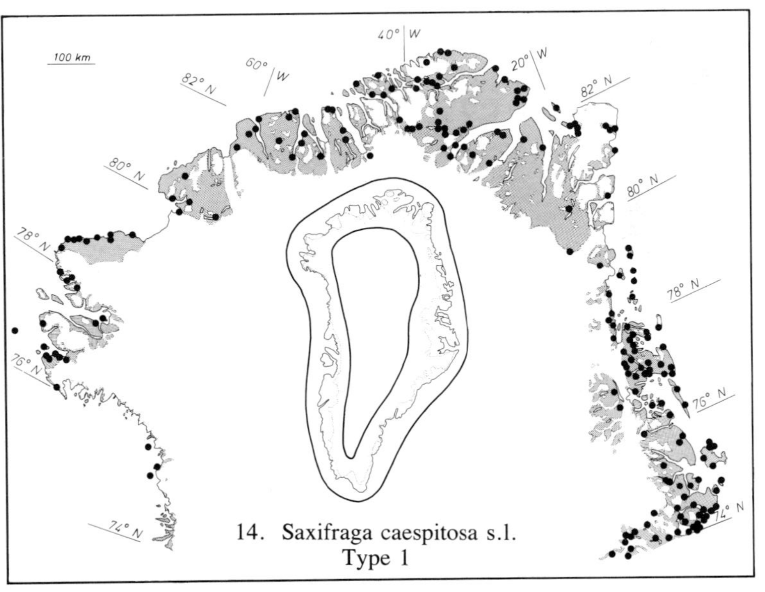

14. Saxifraga caespitosa s.l.
Type 1

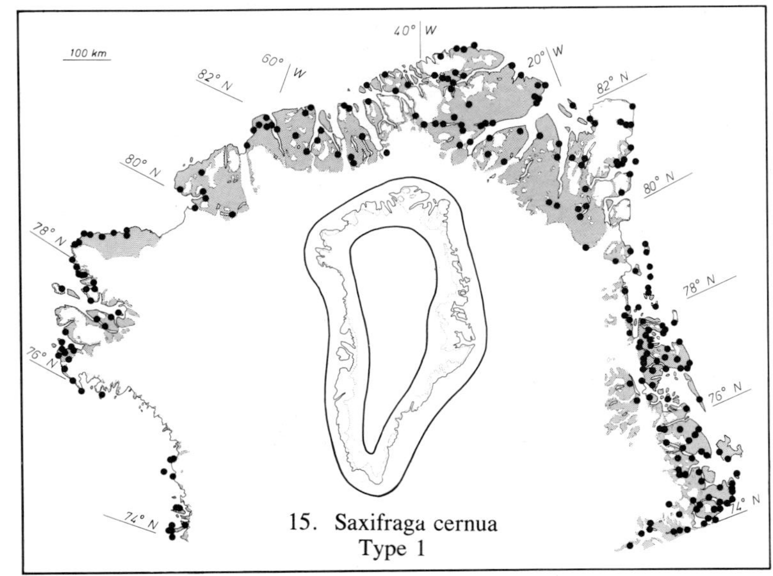

15. Saxifraga cernua
Type 1

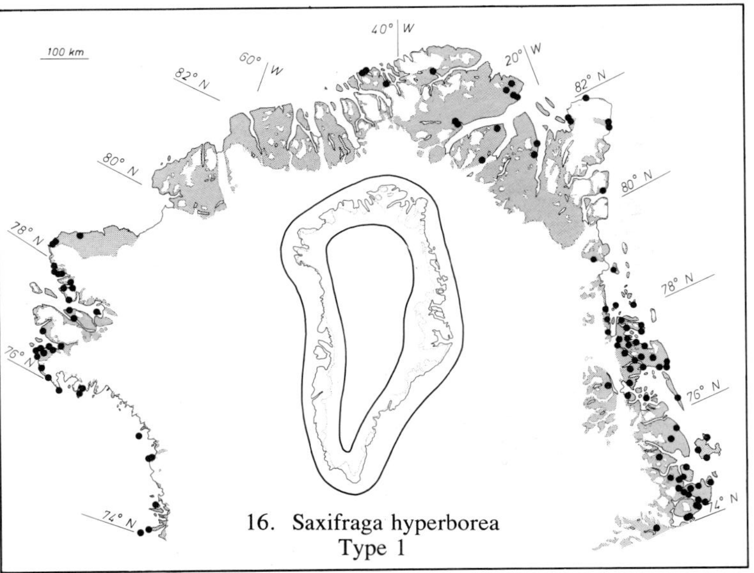

16. Saxifraga hyperborea
Type 1

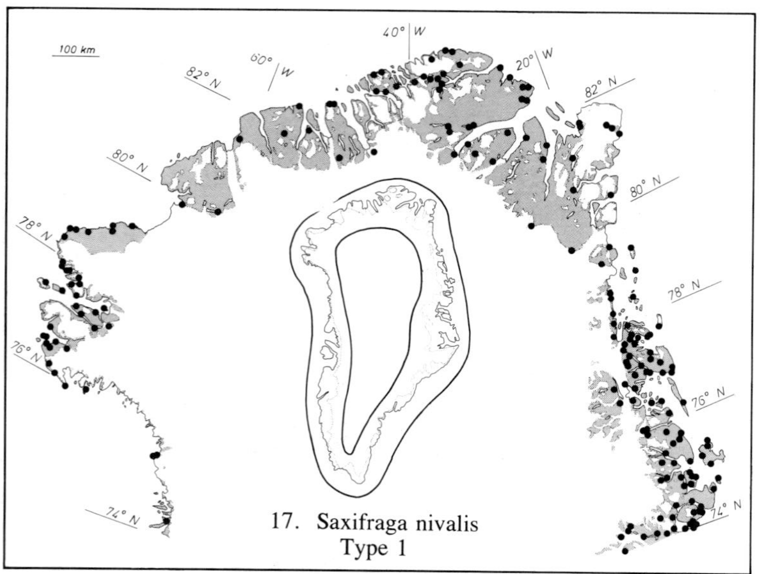

17. Saxifraga nivalis
Type 1

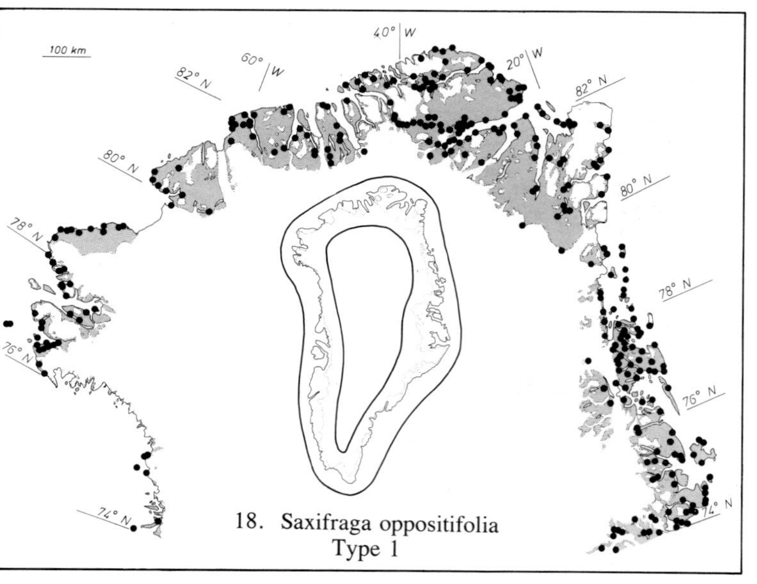

18. Saxifraga oppositifolia
Type 1

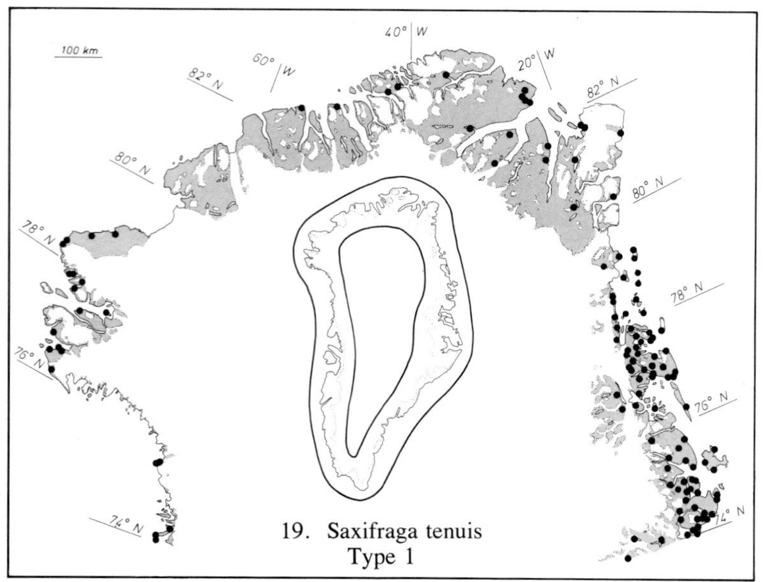

19. Saxifraga tenuis
Type 1

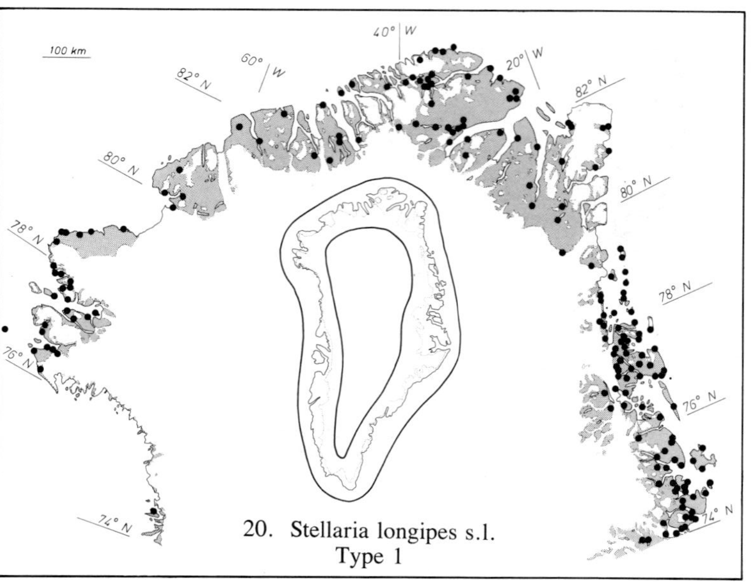

20. Stellaria longipes s.l.
Type 1

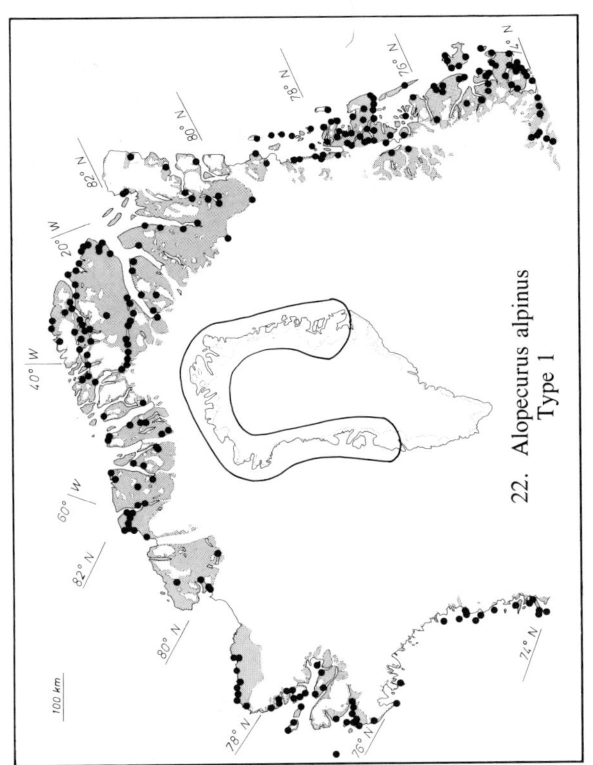

22. Alopecurus alpinus
Type 1

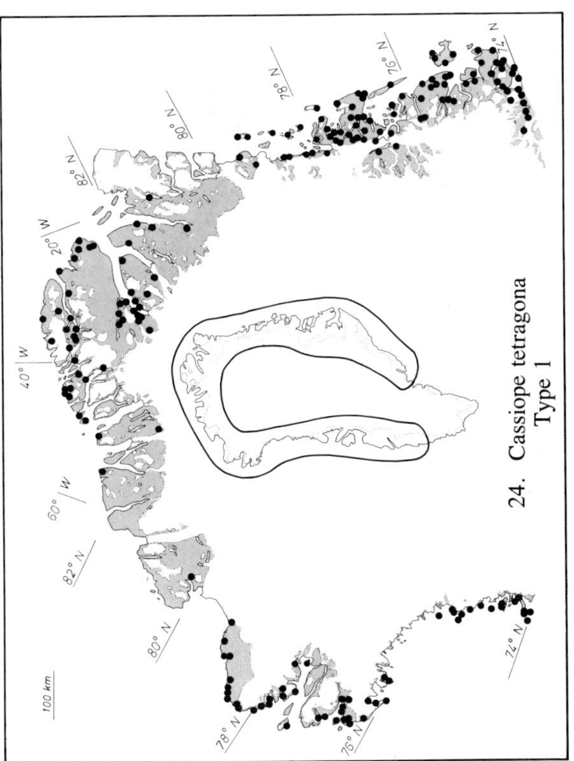

24. Cassiope tetragona
Type 1

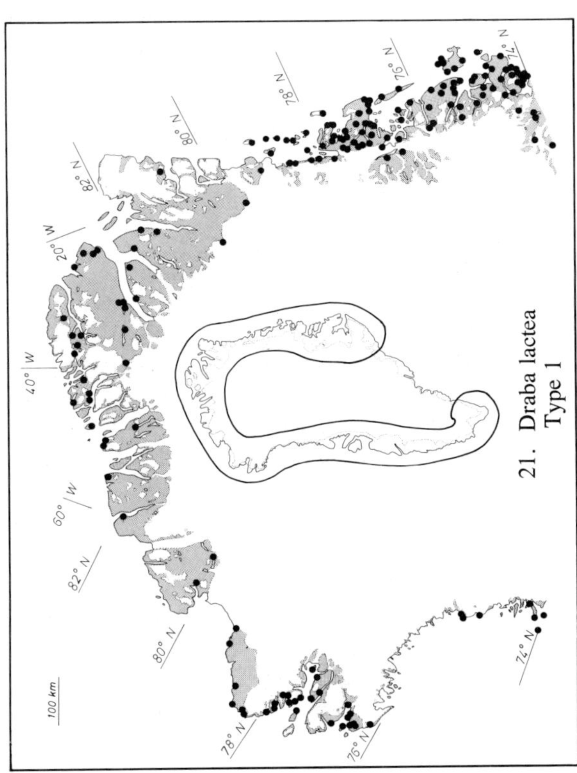

21. Draba lactea
Type 1

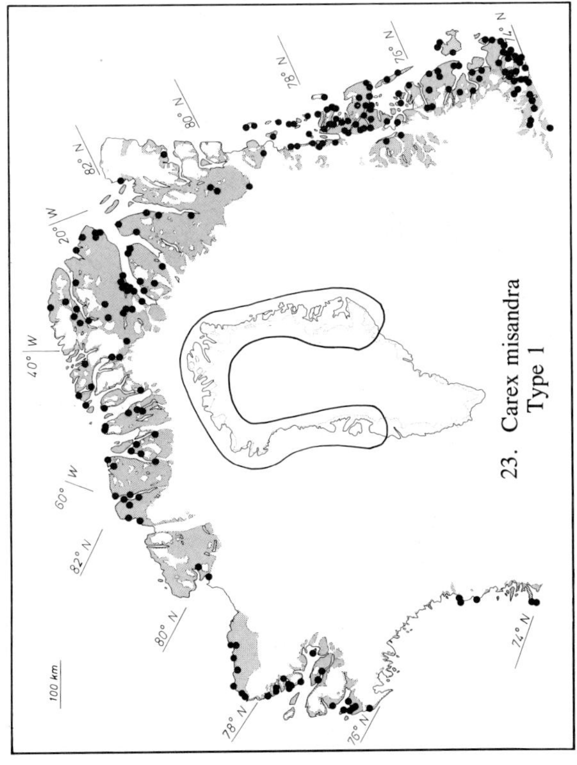

23. Carex misandra
Type 1

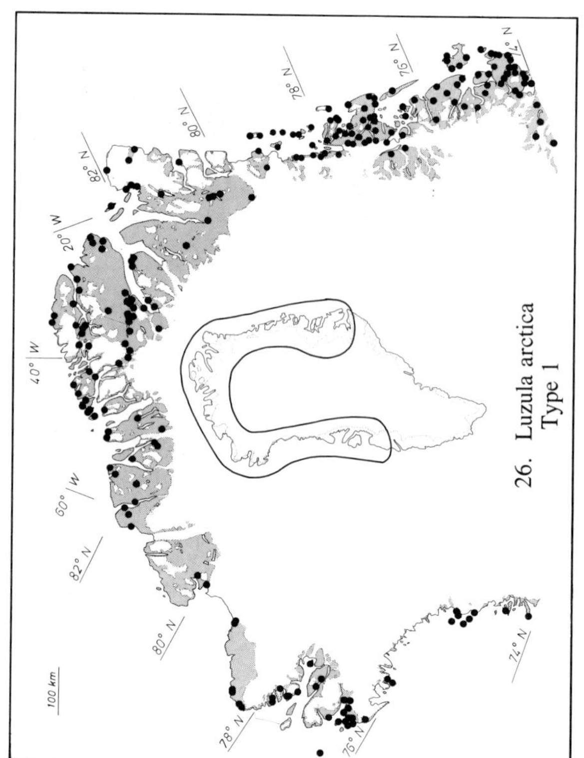

26. Luzula arctica
Type 1

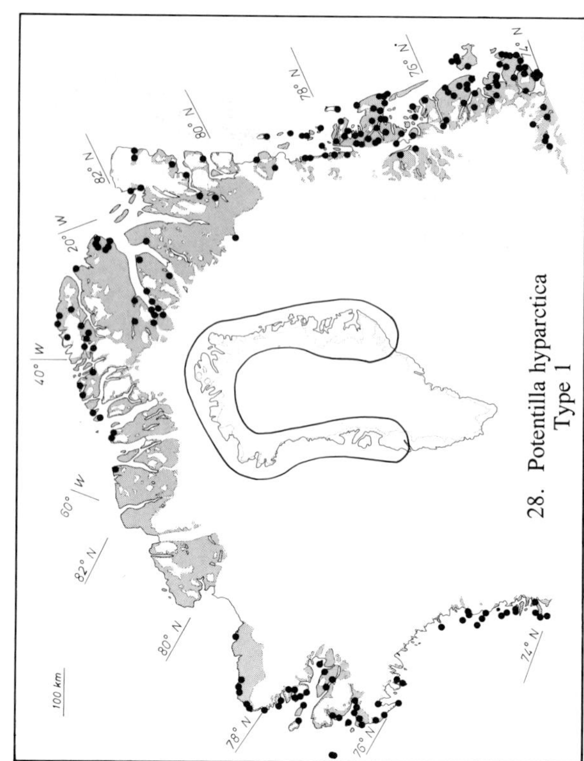

28. Potentilla hyparctica
Type 1

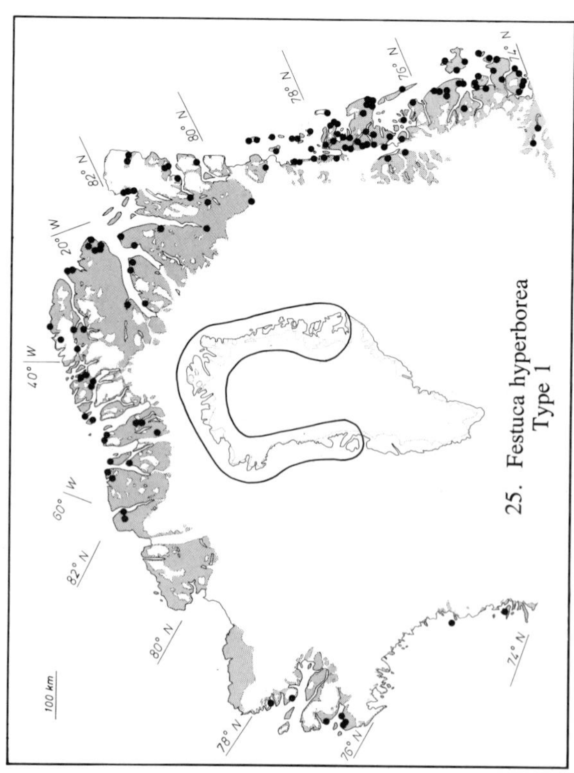

25. Festuca hyperborea
Type 1

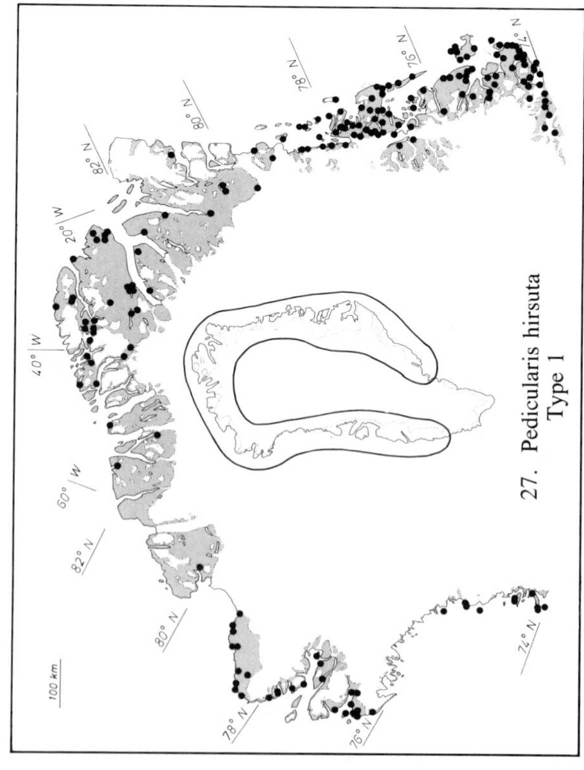

27. Pedicularis hirsuta
Type 1

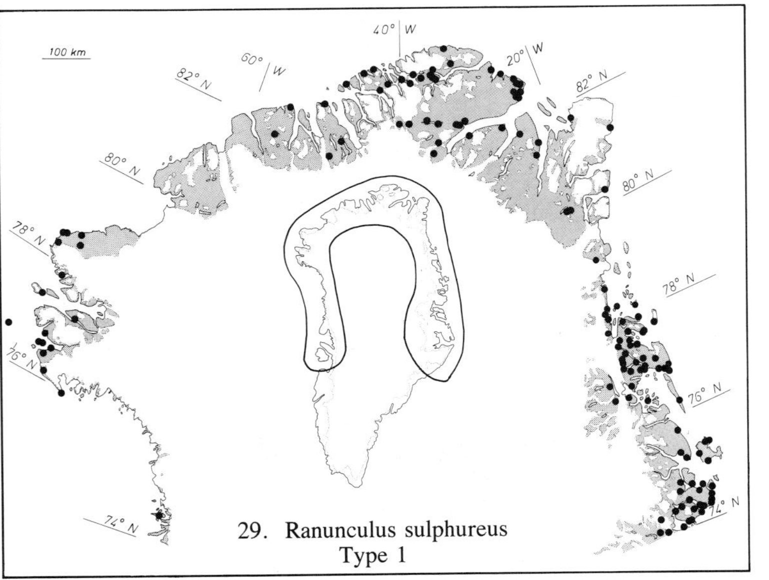

29. Ranunculus sulphureus
Type 1

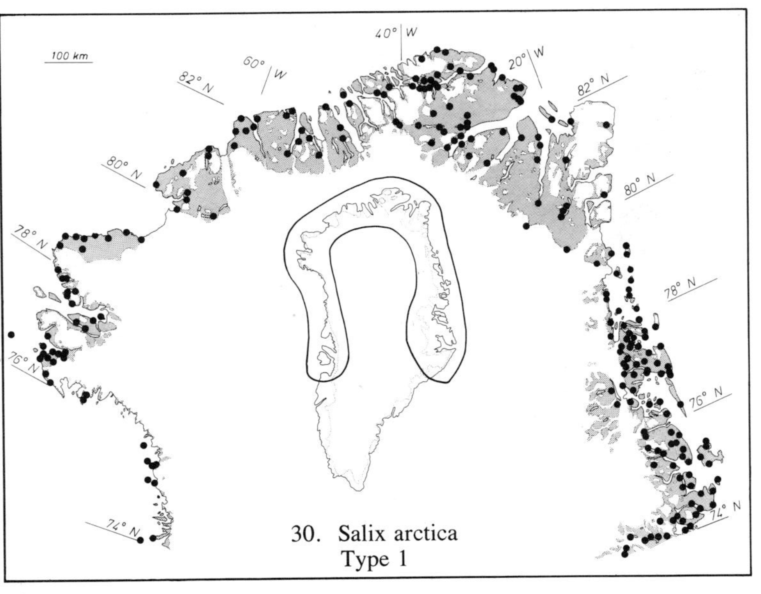

30. Salix arctica
Type 1

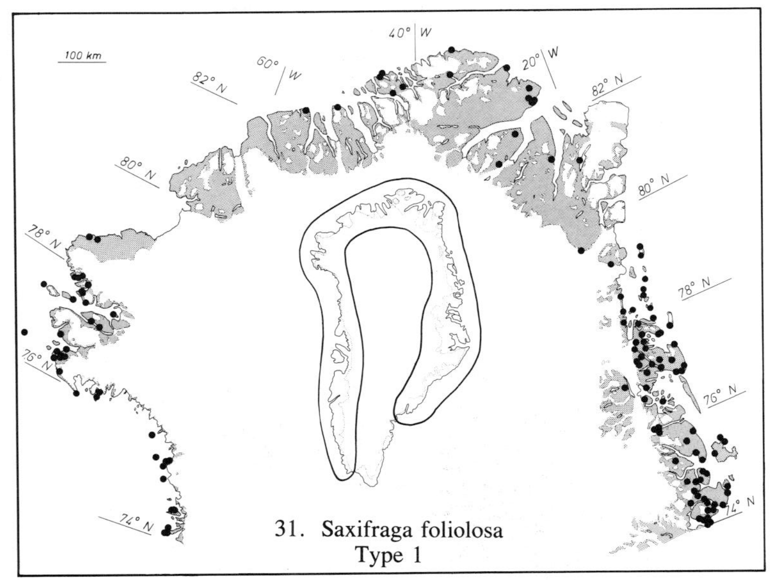

31. Saxifraga foliolosa
Type 1

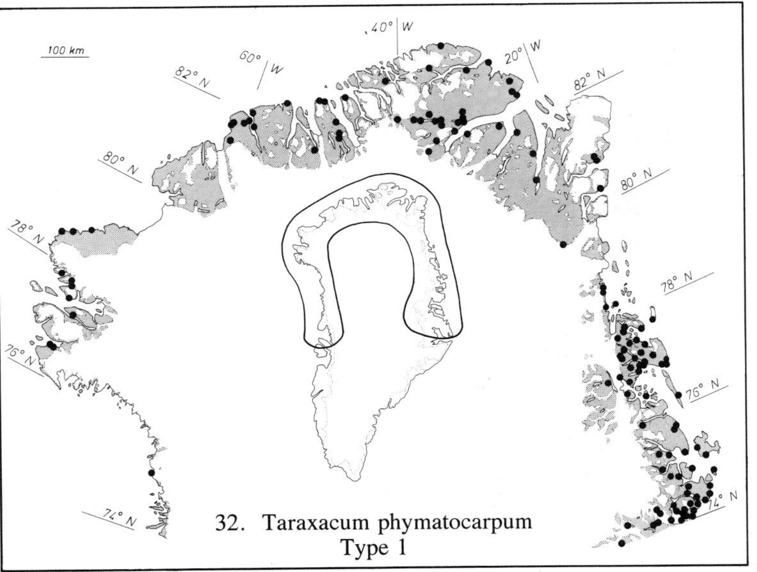

32. Taraxacum phymatocarpum
Type 1

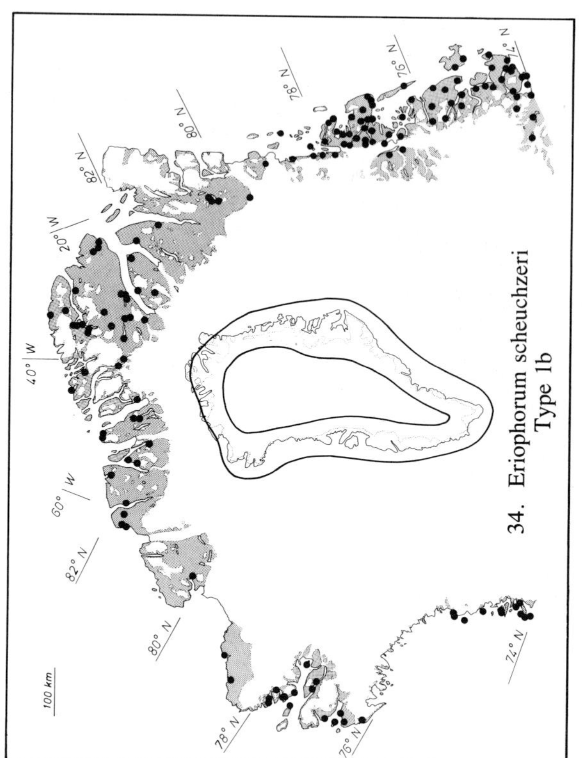

34. Eriophorum scheuchzeri
Type 1b

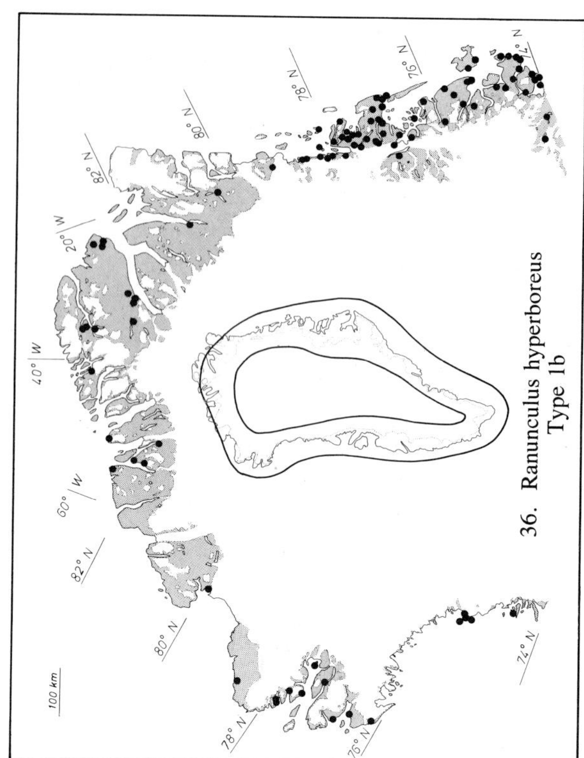

36. Ranunculus hyperboreus
Type 1b

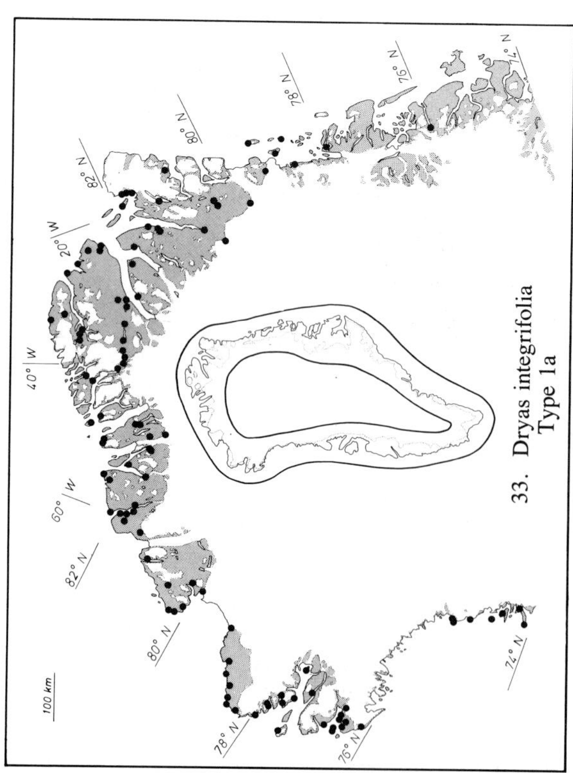

33. Dryas integrifolia
Type 1a

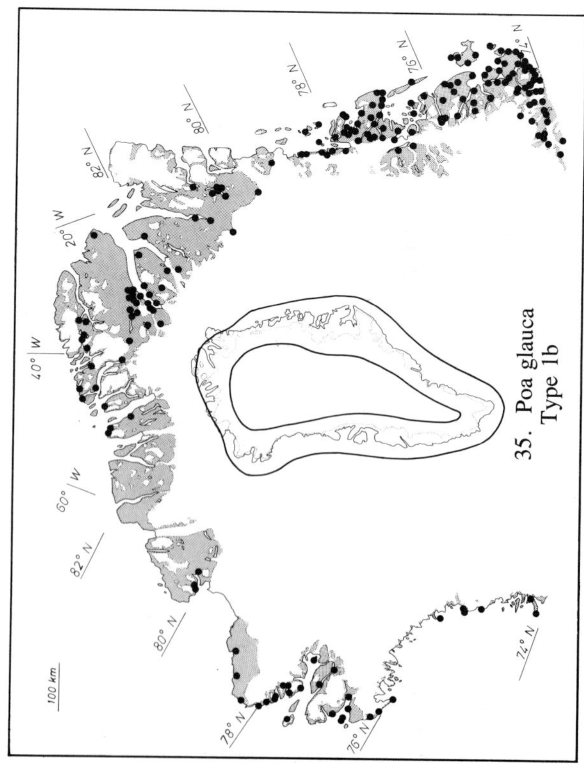

35. Poa glauca
Type 1b

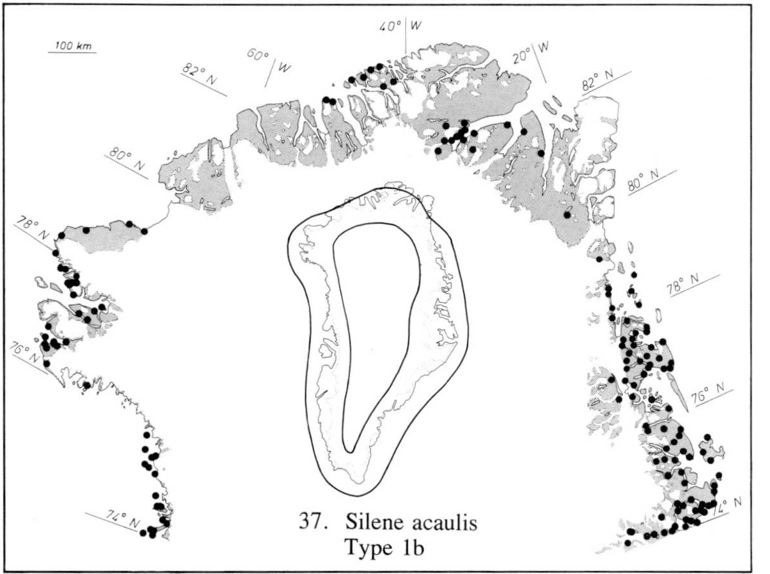

37. Silene acaulis
Type 1b

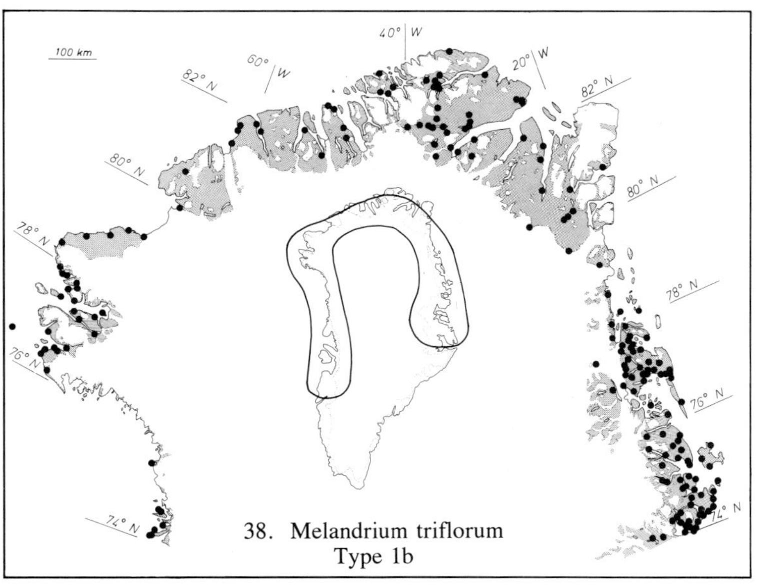

38. Melandrium triflorum
Type 1b

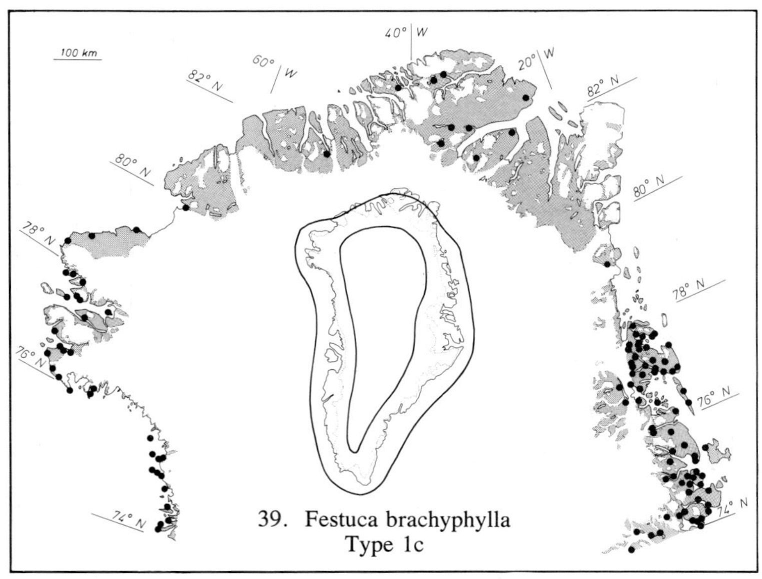

39. Festuca brachyphylla
Type 1c

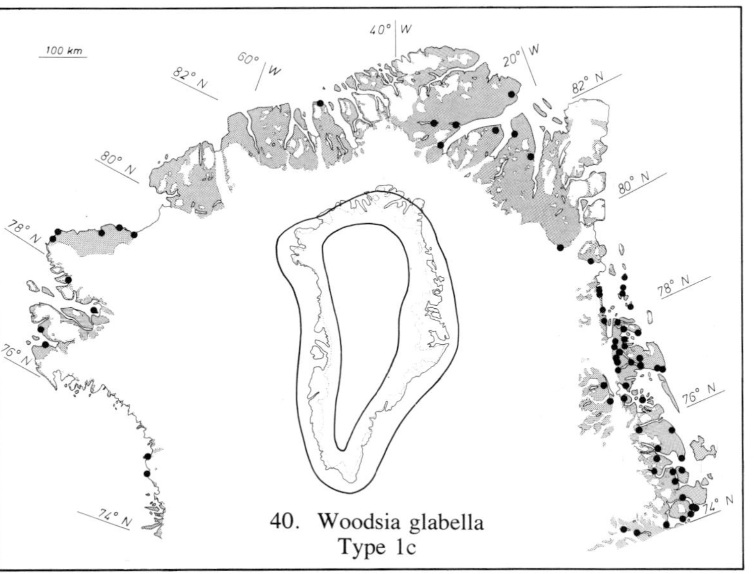

40. Woodsia glabella
Type 1c

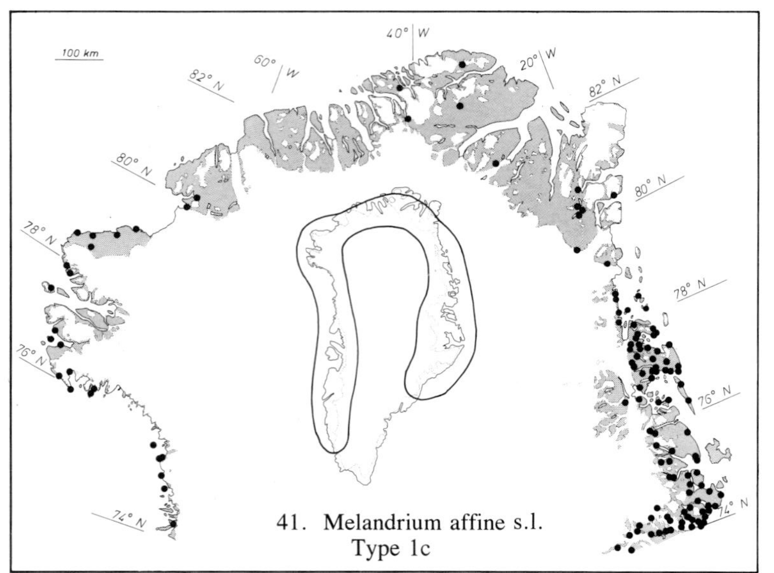

41. Melandrium affine s.l.
Type 1c

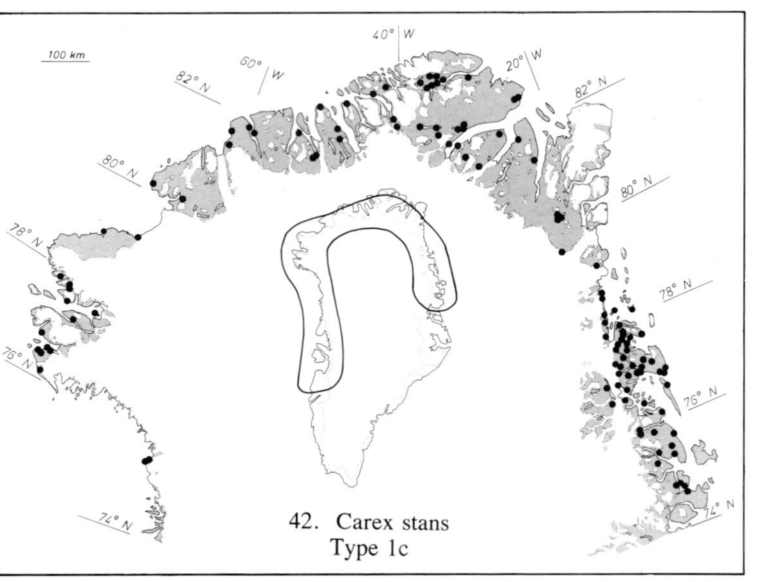

42. Carex stans
Type 1c

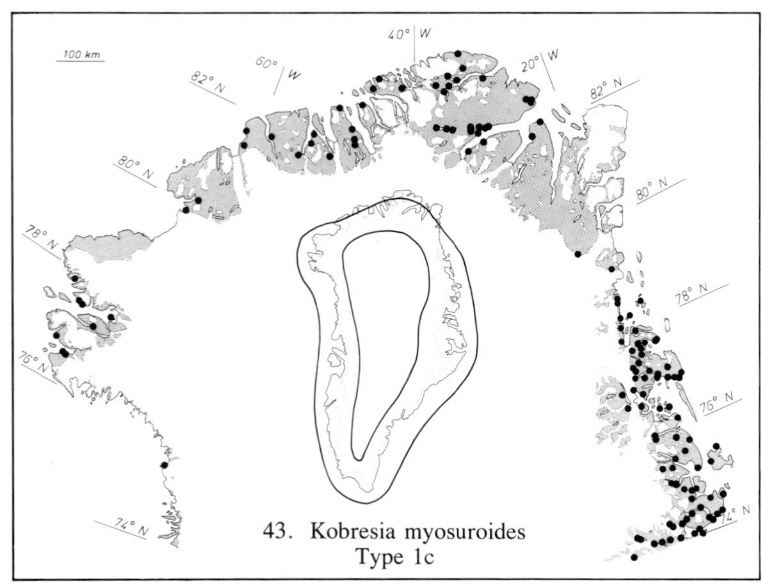

43. Kobresia myosuroides
Type 1c

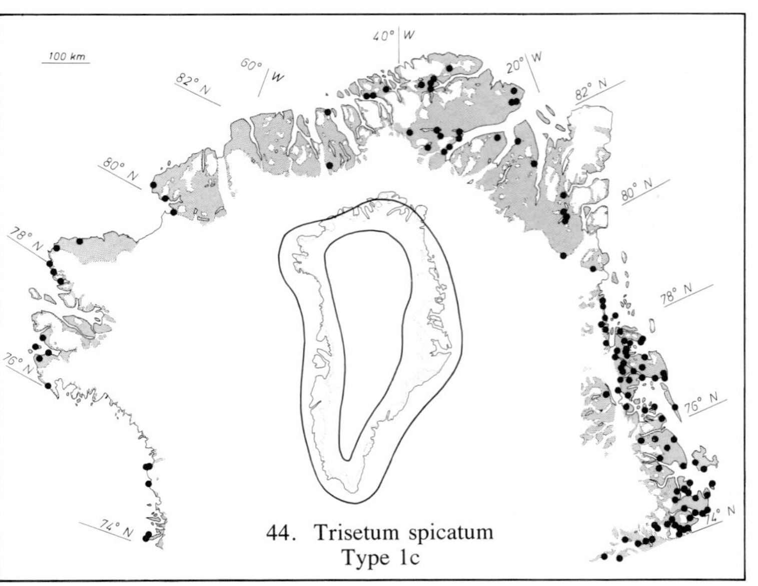

44. Trisetum spicatum
Type 1c

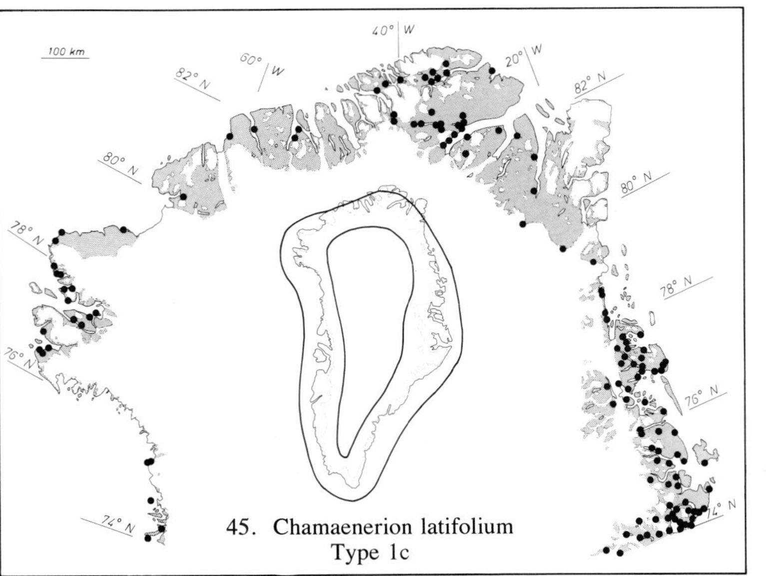

45. Chamaenerion latifolium
Type 1c

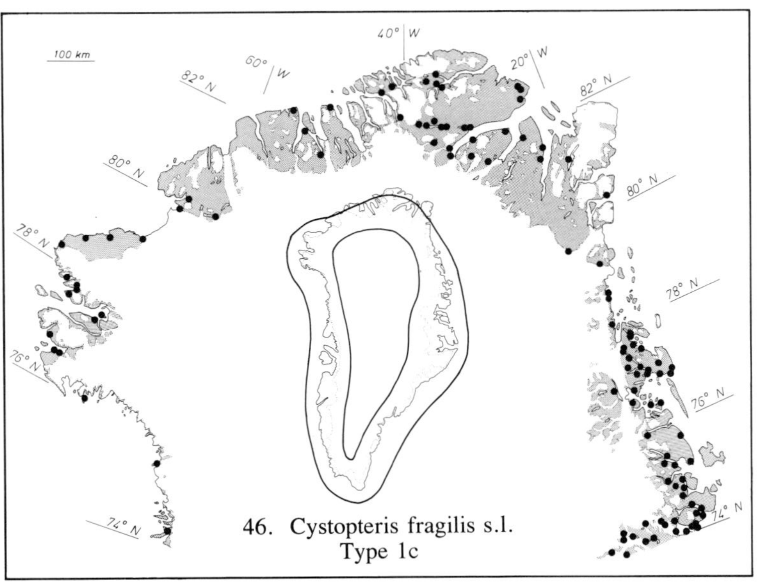

46. Cystopteris fragilis s.l.
Type 1c

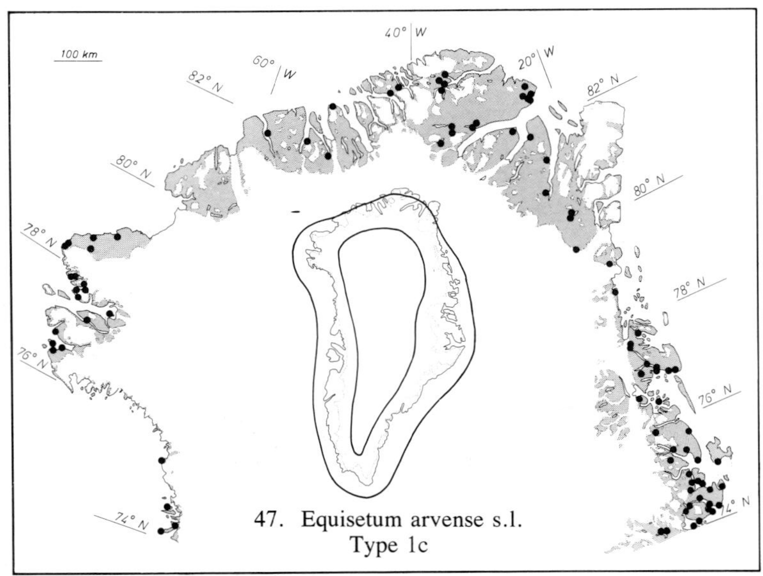

47. Equisetum arvense s.l.
Type 1c

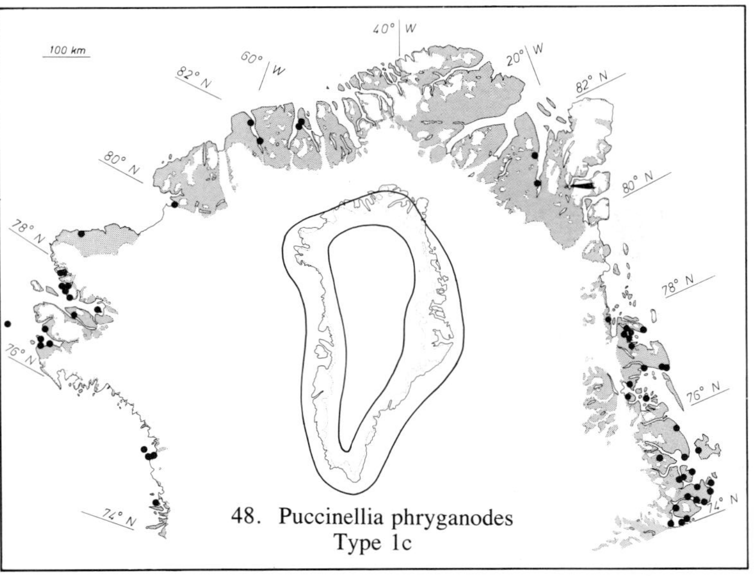

48. Puccinellia phryganodes
Type 1c

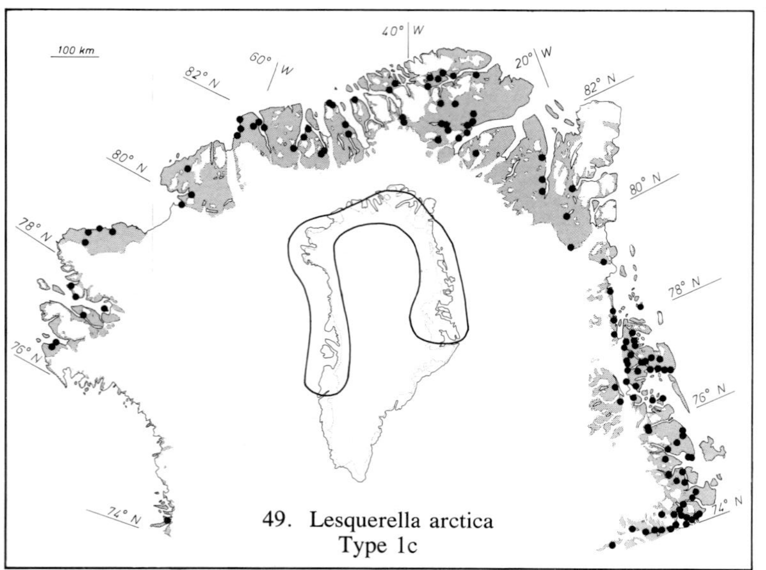

49. Lesquerella arctica
Type 1c

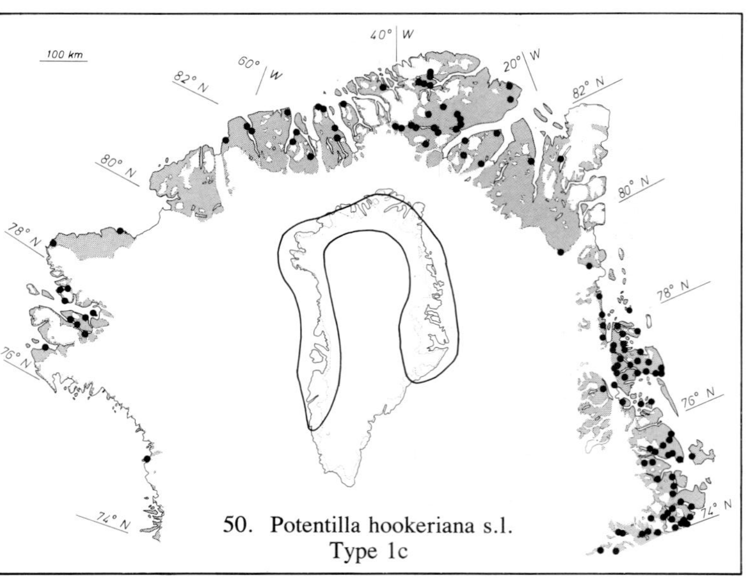

50. Potentilla hookeriana s.l.
Type 1c

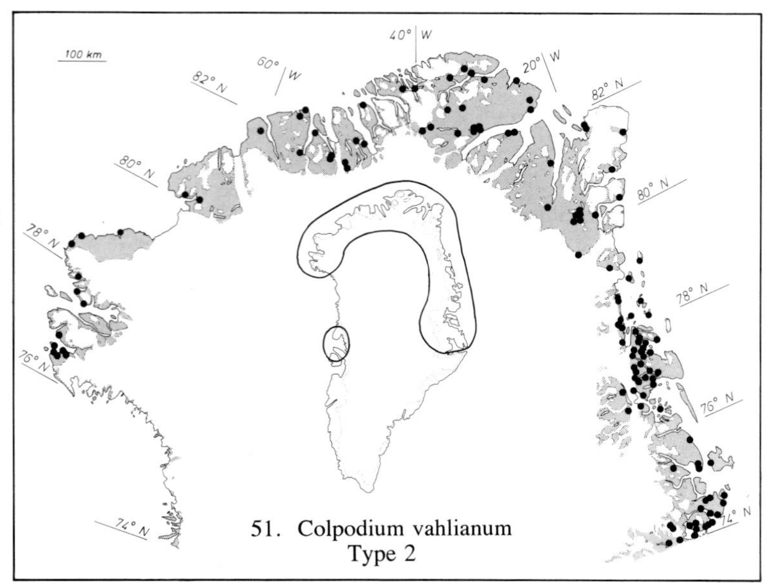

51. Colpodium vahlianum
Type 2

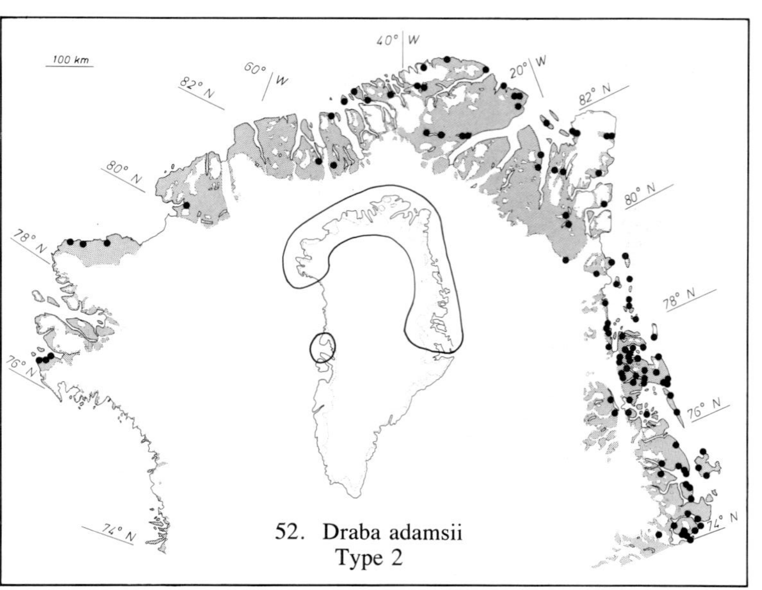

52. Draba adamsii
Type 2

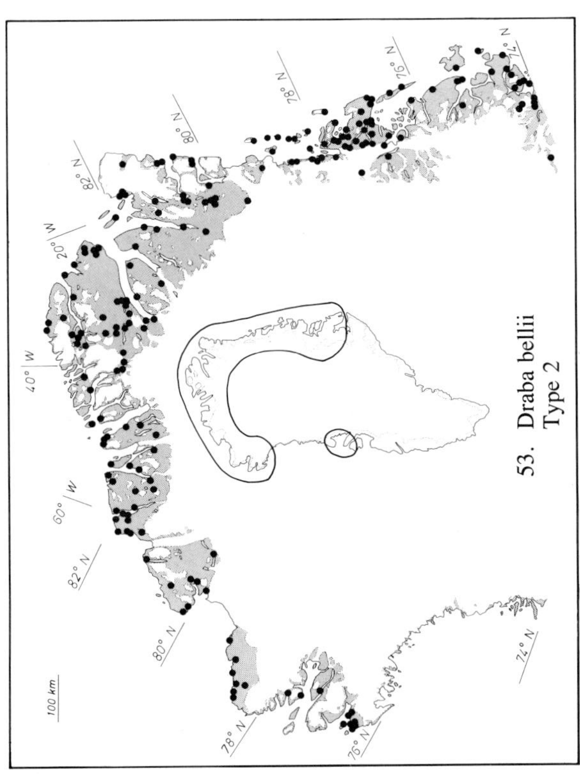

53. Draba bellii
Type 2

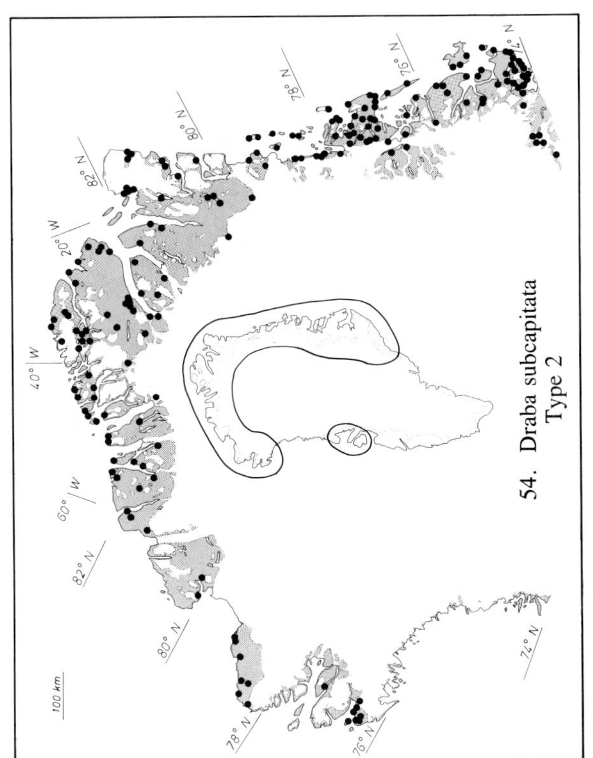

54. Draba subcapitata
Type 2

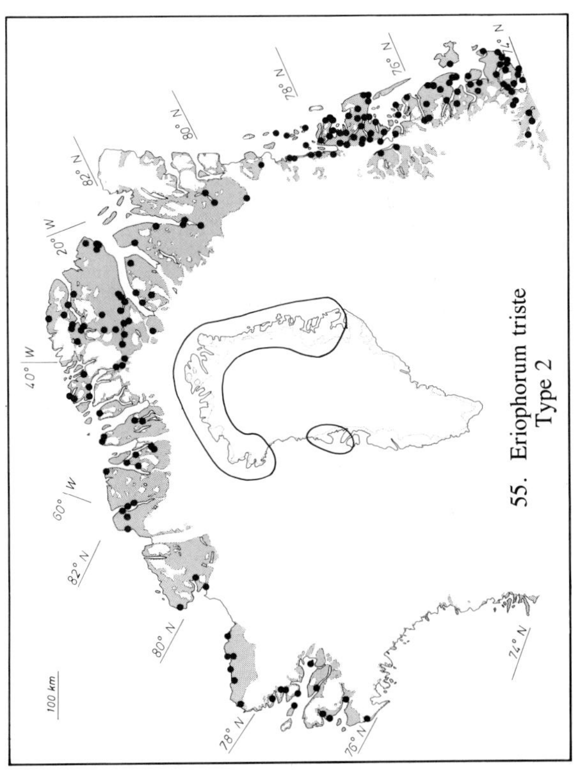

55. Eriophorum triste
Type 2

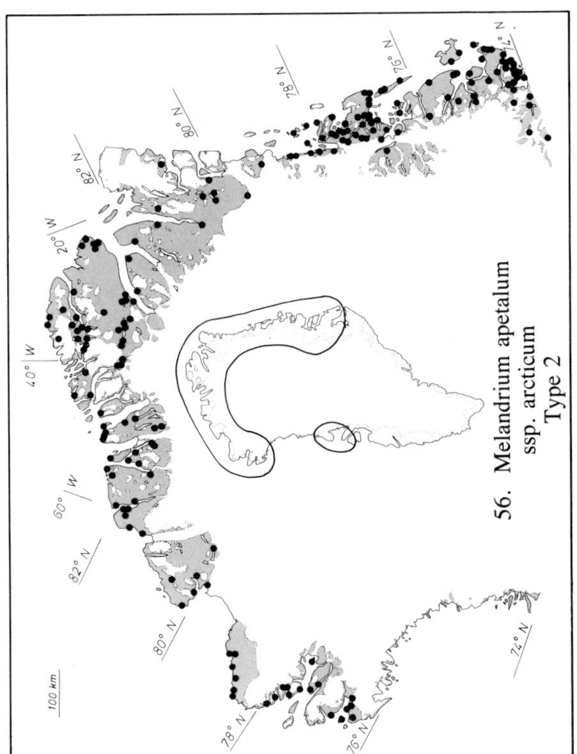

56. Melandrium apetalum
ssp. arcticum
Type 2

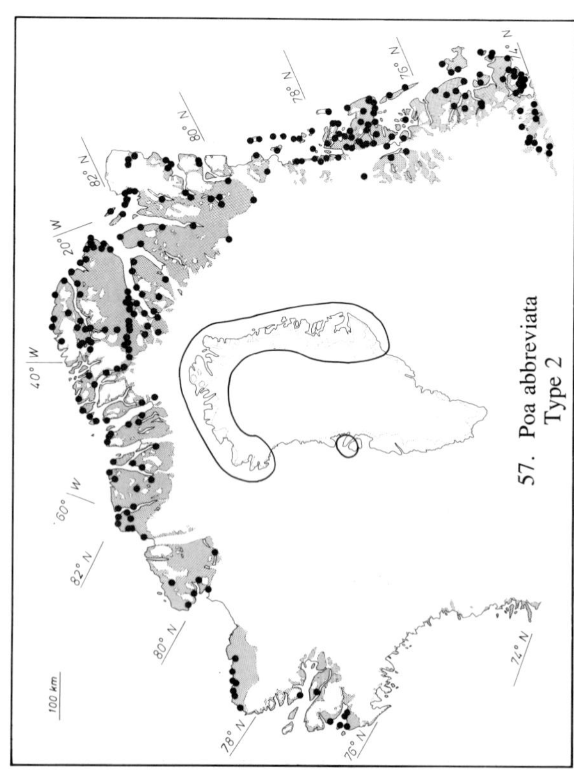

57. Poa abbreviata
Type 2

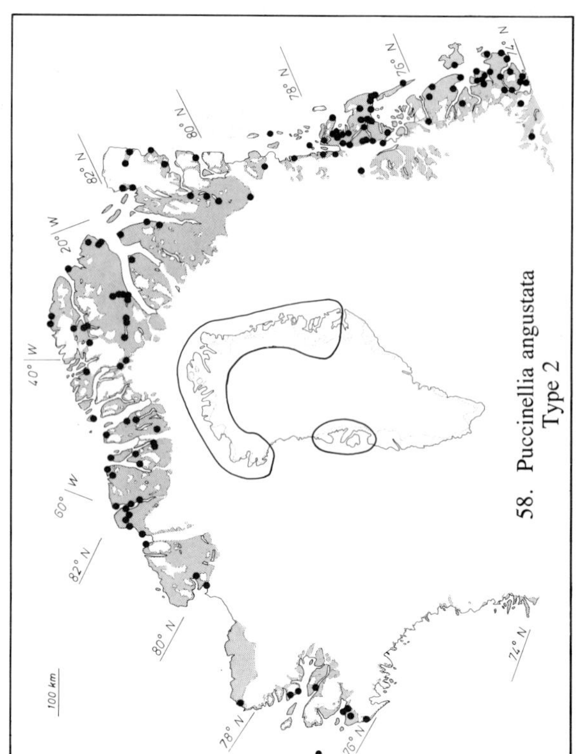

58. Puccinellia angustata
Type 2

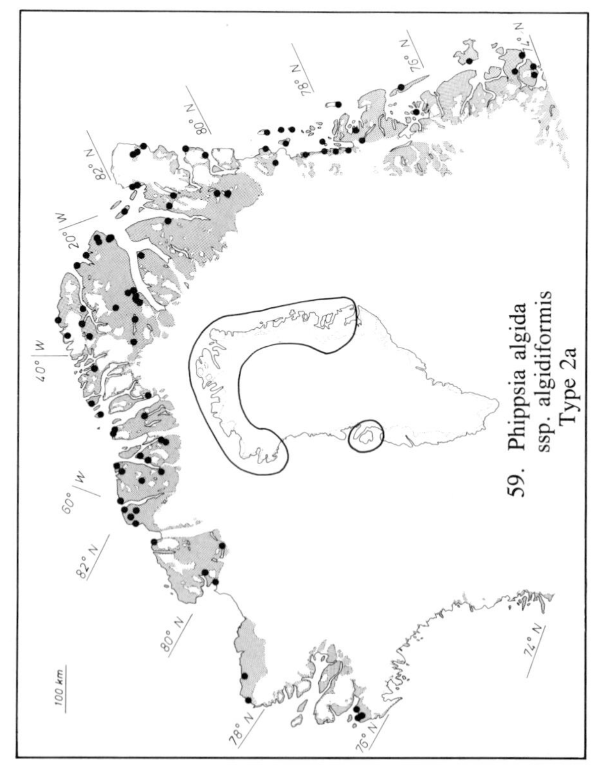

59. Phippsia algida
ssp. algidiformis
Type 2a

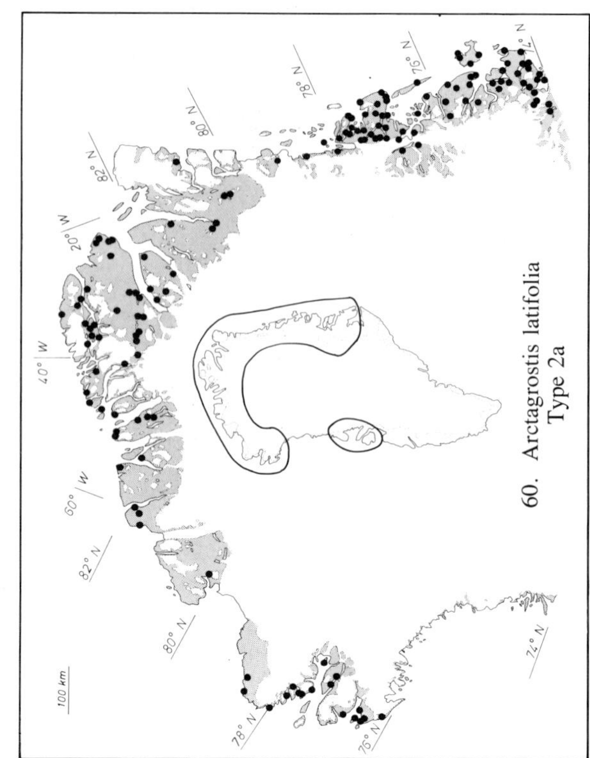

60. Arctagrostis latifolia
Type 2a

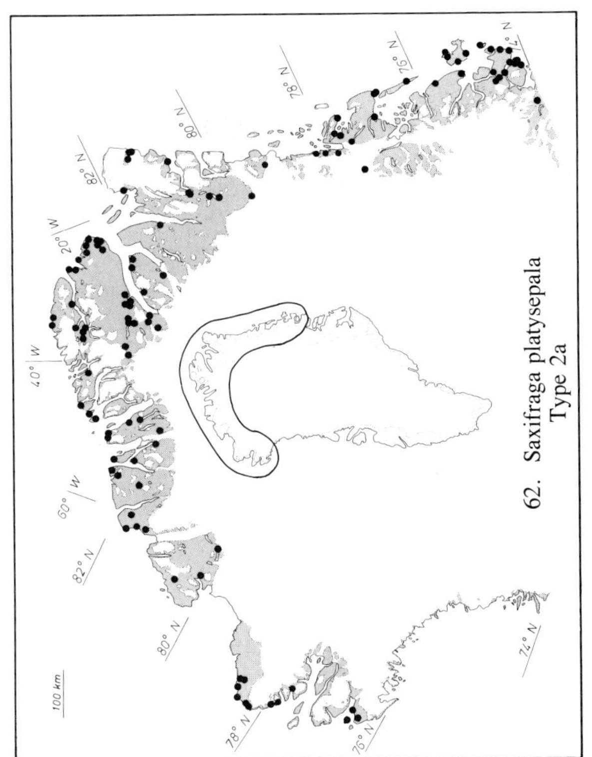

62. Saxifraga platysepala
Type 2a

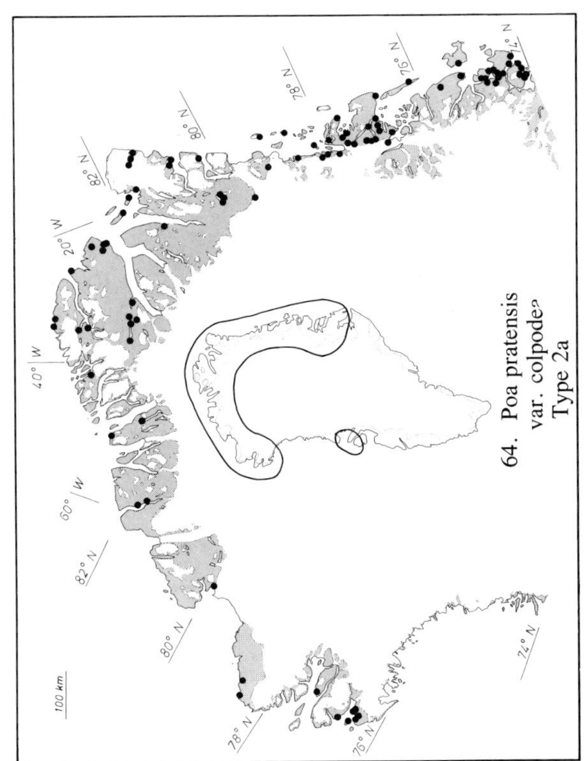

64. Poa pratensis
var. colpodea
Type 2a

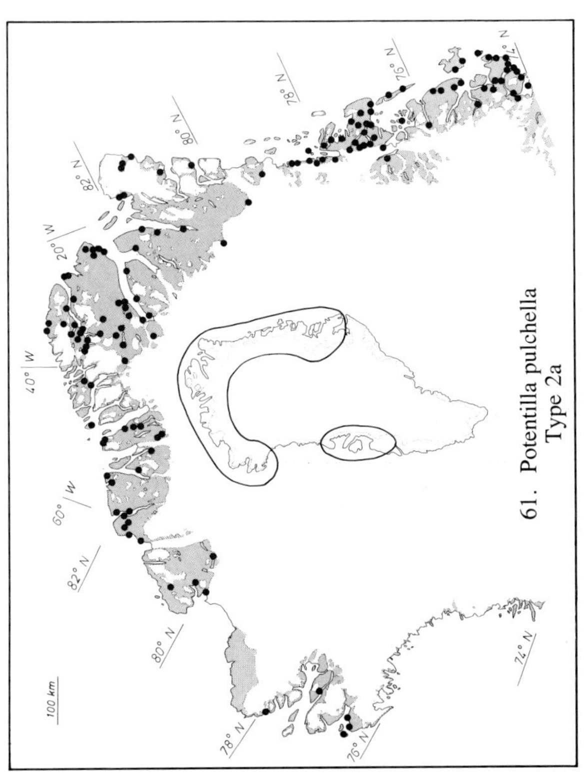

61. Potentilla pulchella
Type 2a

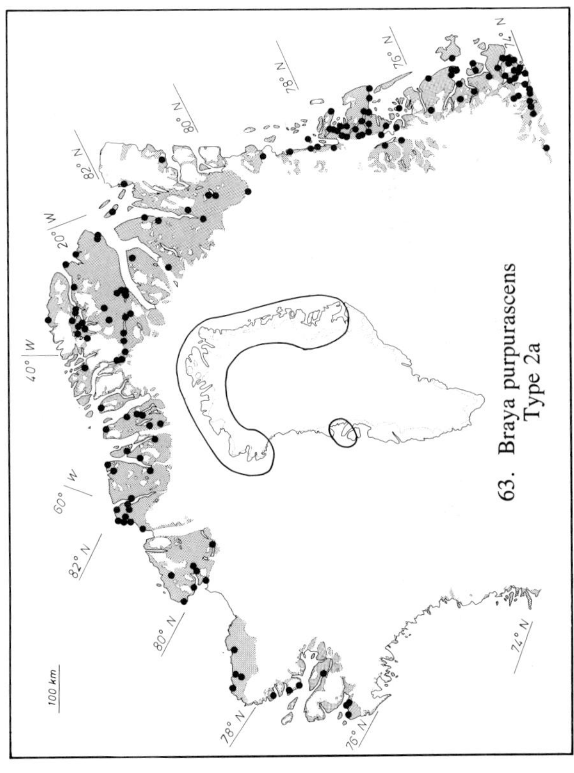

63. Braya purpurascens
Type 2a

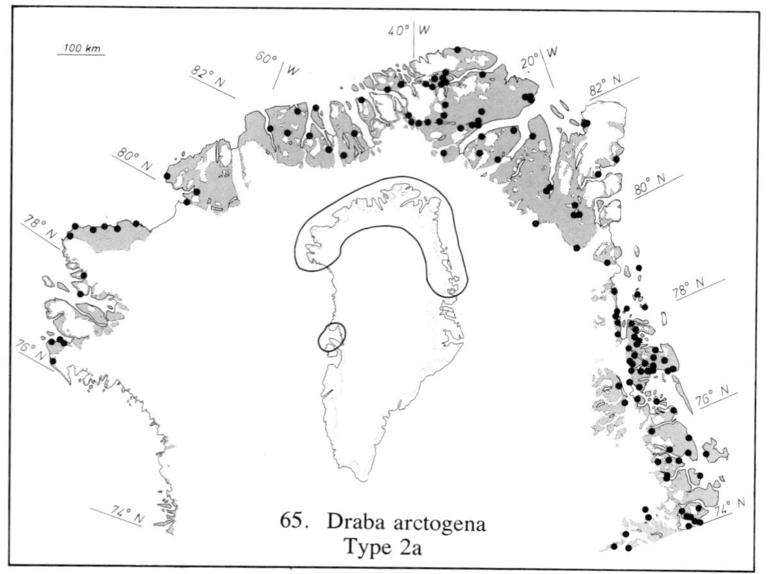

65. Draba arctogena
Type 2a

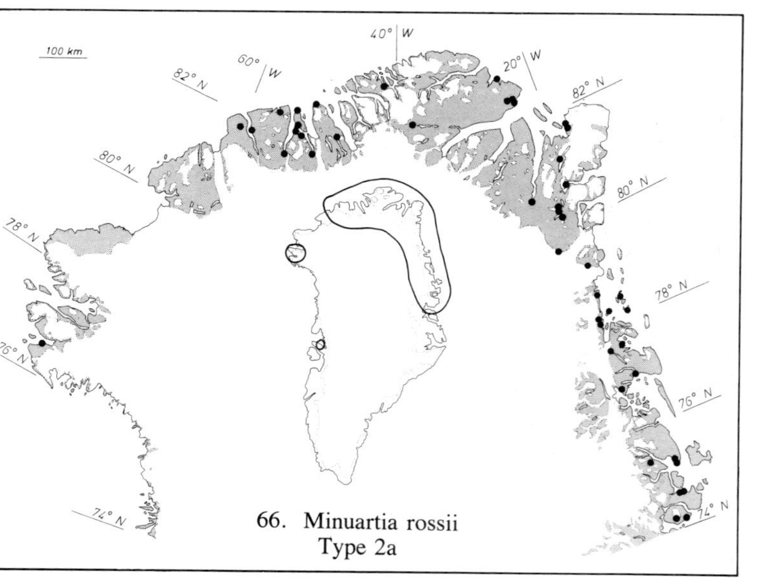

66. Minuartia rossii
Type 2a

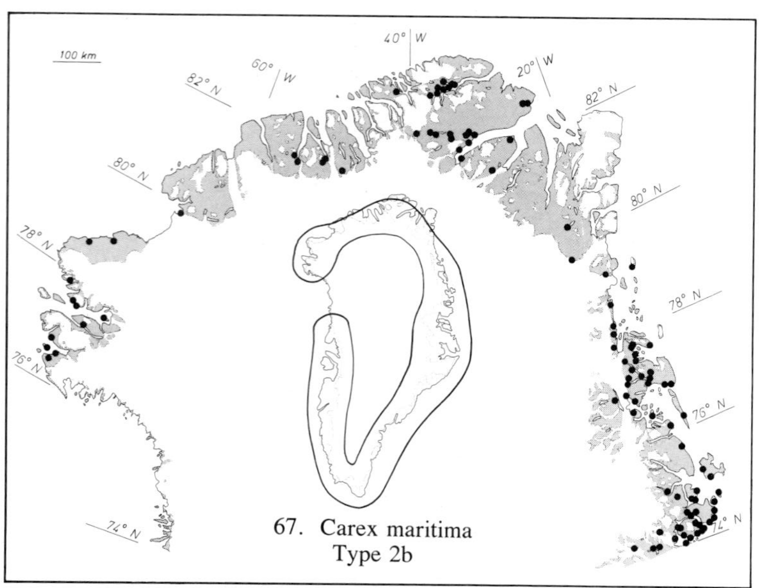

67. Carex maritima
Type 2b

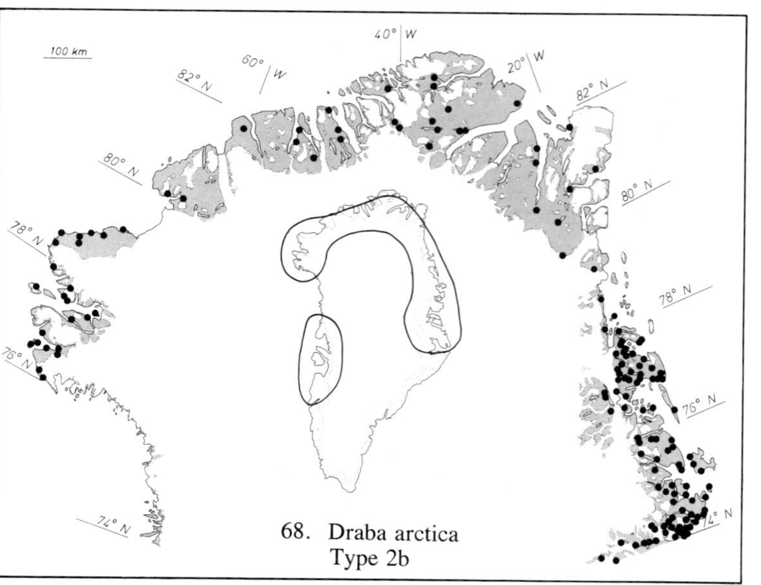

68. Draba arctica
Type 2b

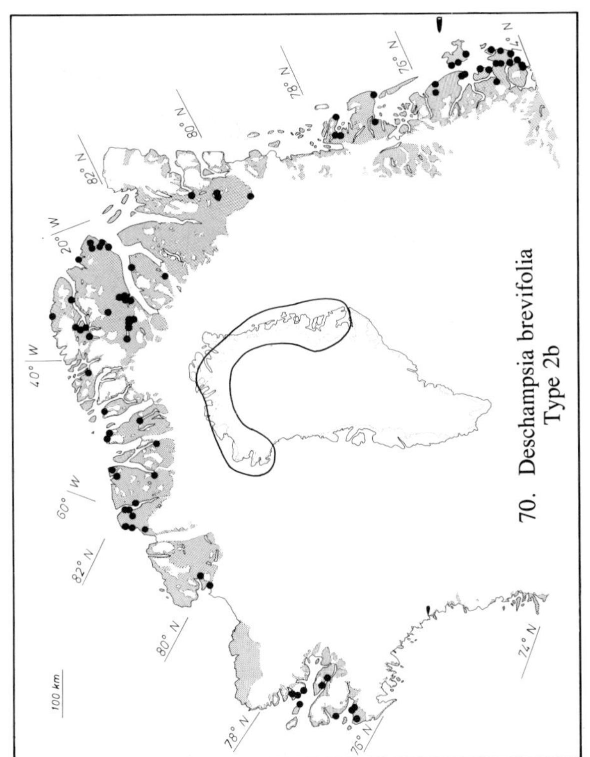
70. Deschampsia brevifolia
Type 2b

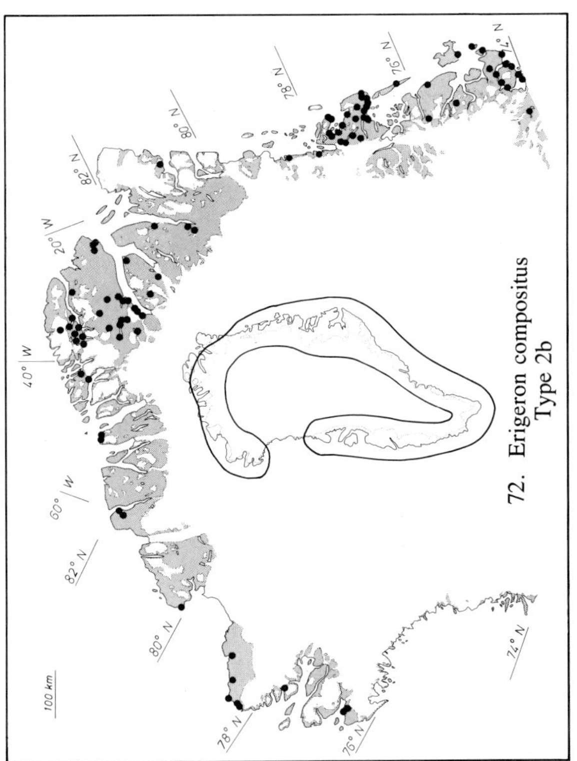
72. Erigeron compositus
Type 2b

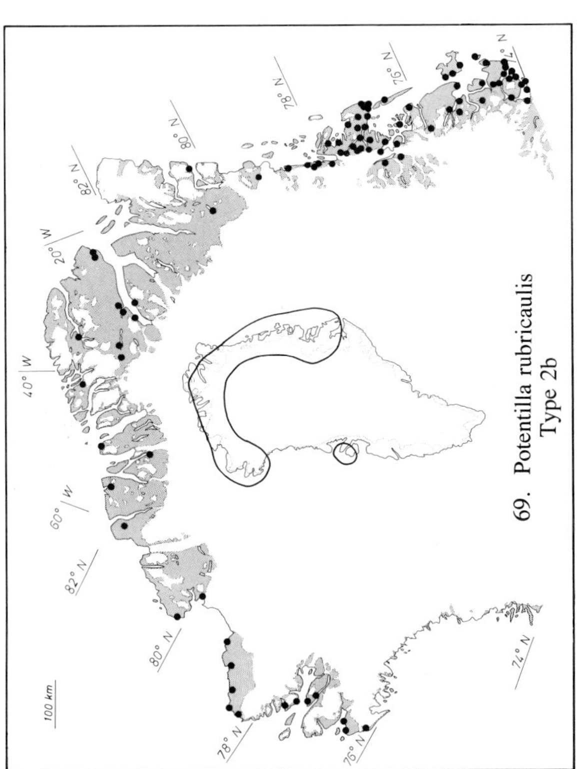
69. Potentilla rubricaulis
Type 2b

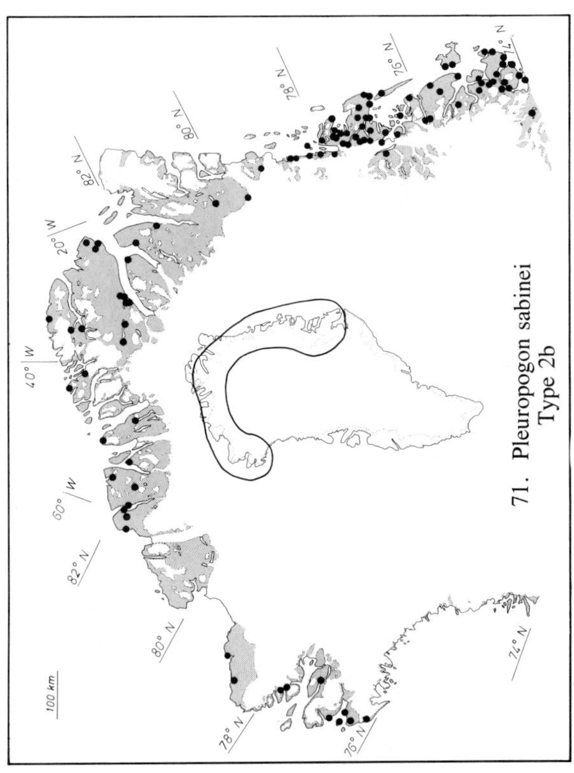
71. Pleuropogon sabinei
Type 2b

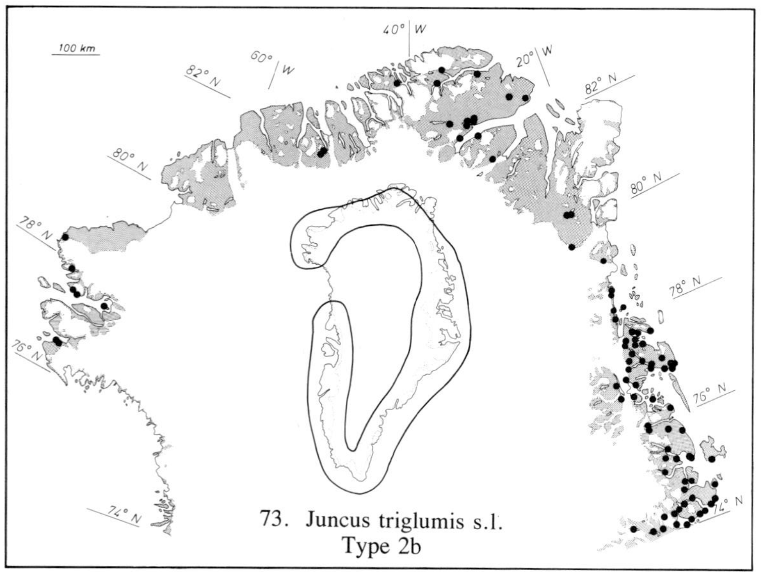

73. Juncus triglumis s.l.
Type 2b

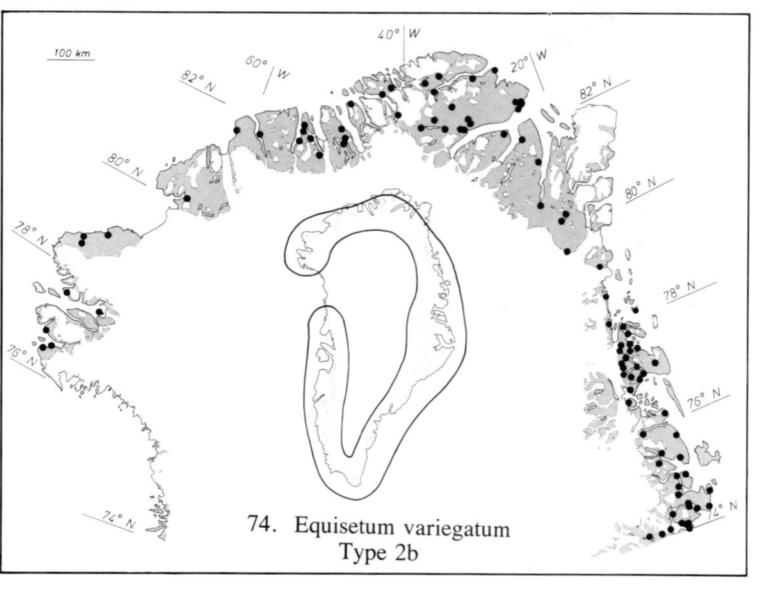

74. Equisetum variegatum
Type 2b

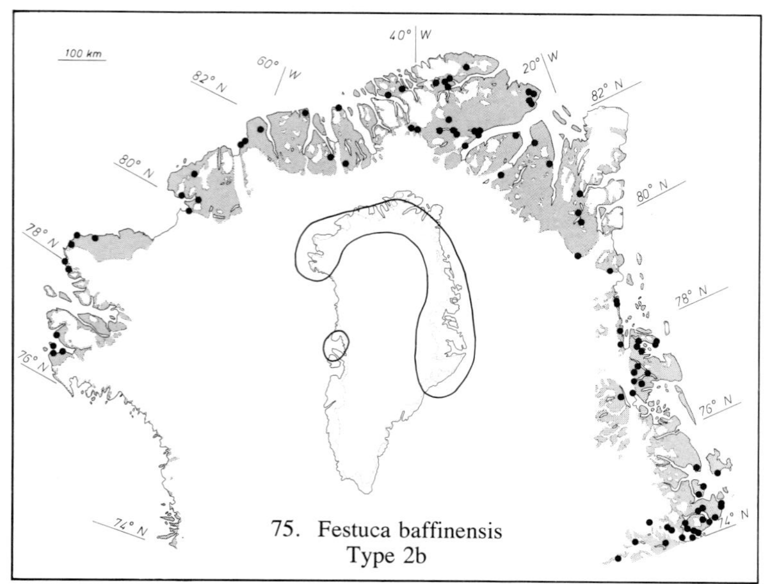

75. Festuca baffinensis
Type 2b

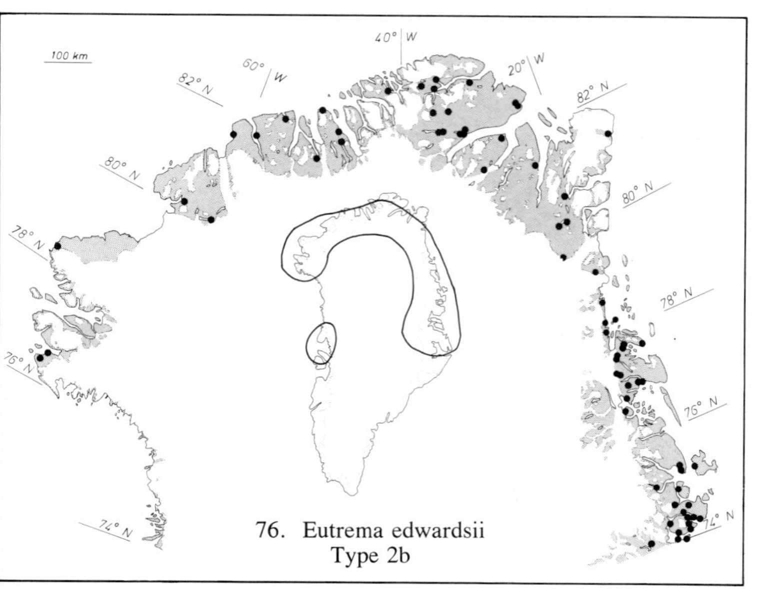

76. Eutrema edwardsii
Type 2b

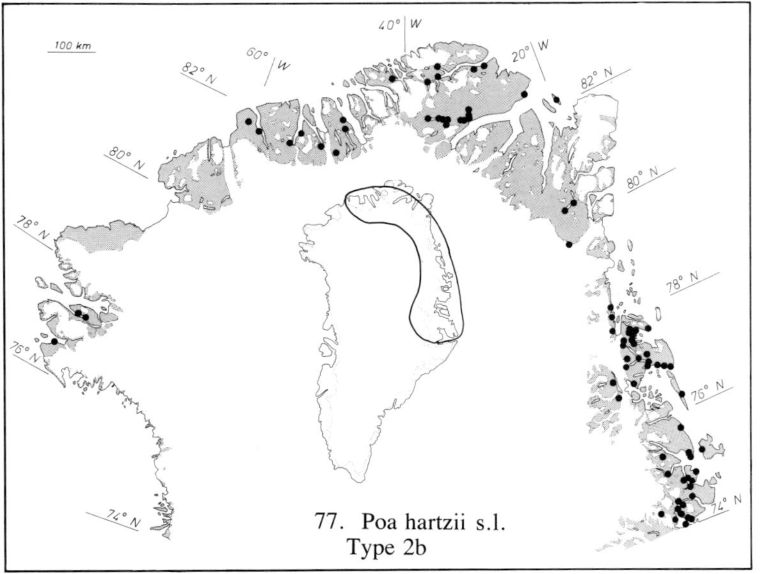

77. Poa hartzii s.l.
Type 2b

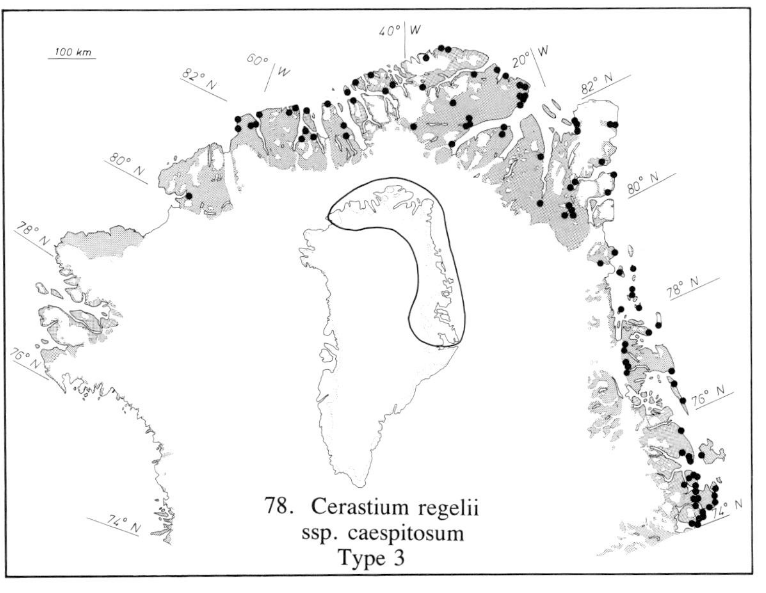

78. Cerastium regelii
ssp. caespitosum
Type 3

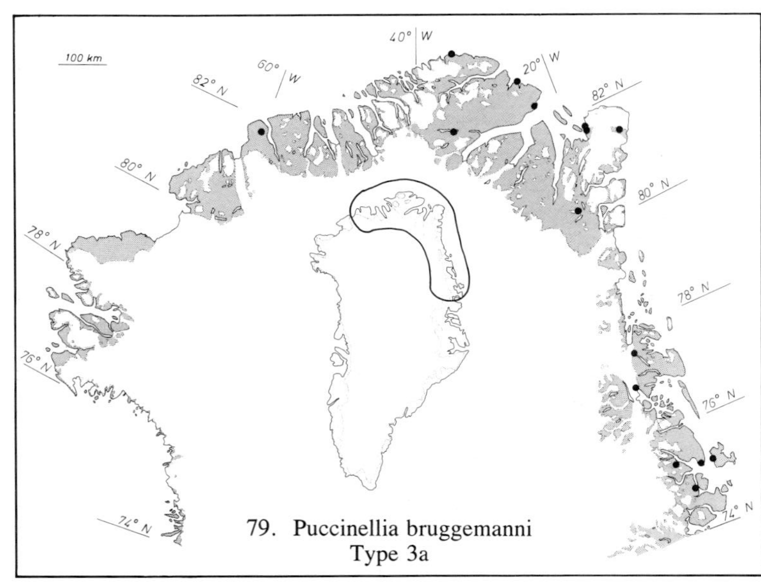

79. Puccinellia bruggemanni
Type 3a

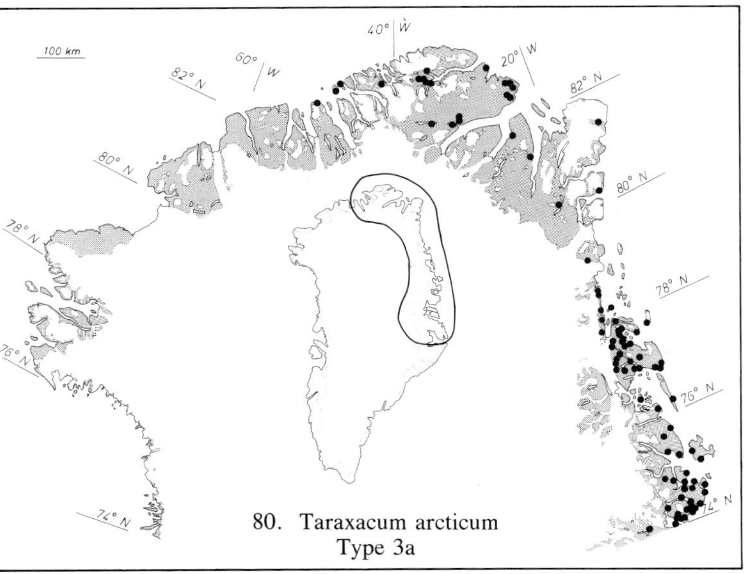

80. Taraxacum arcticum
Type 3a

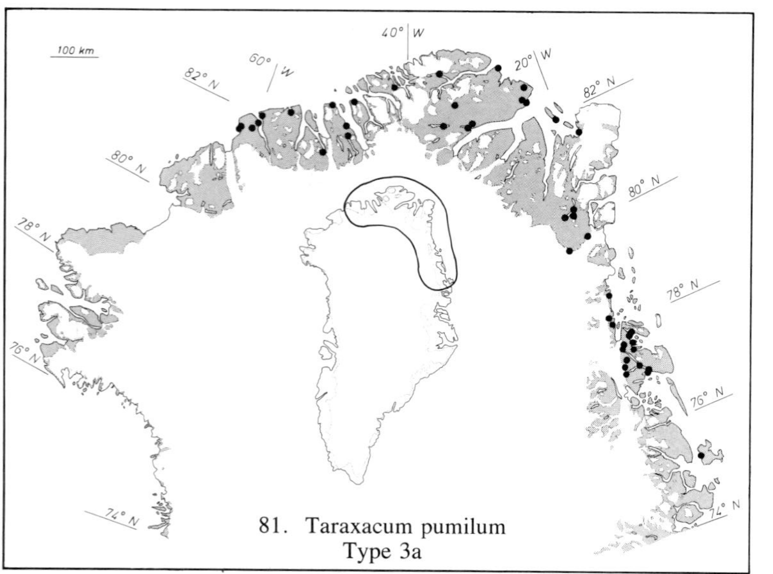

81. Taraxacum pumilum
Type 3a

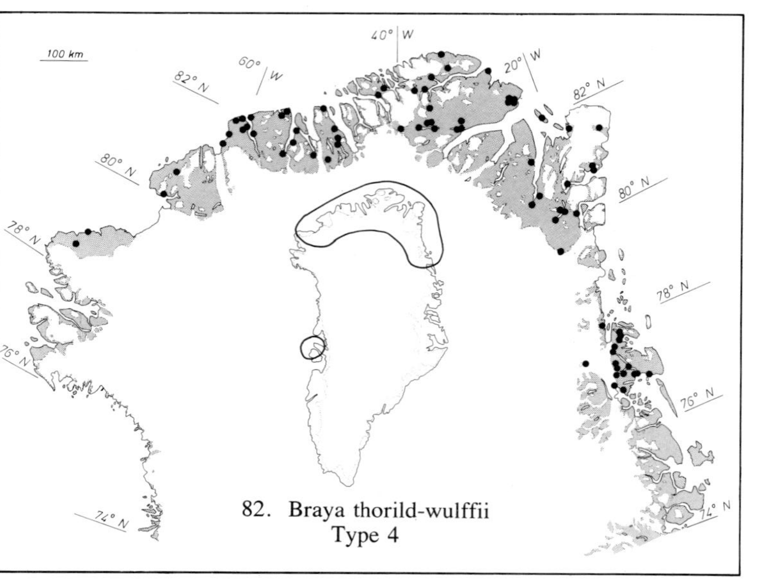

82. Braya thorild-wulffii
Type 4

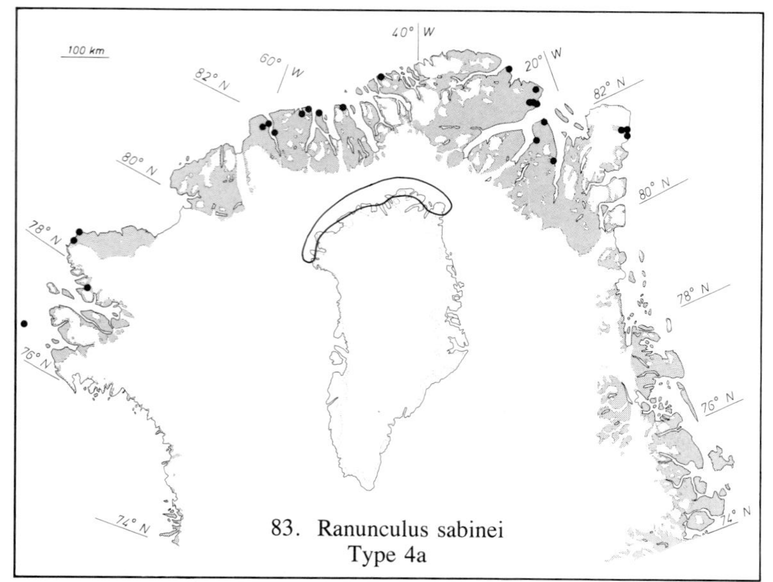

83. Ranunculus sabinei
Type 4a

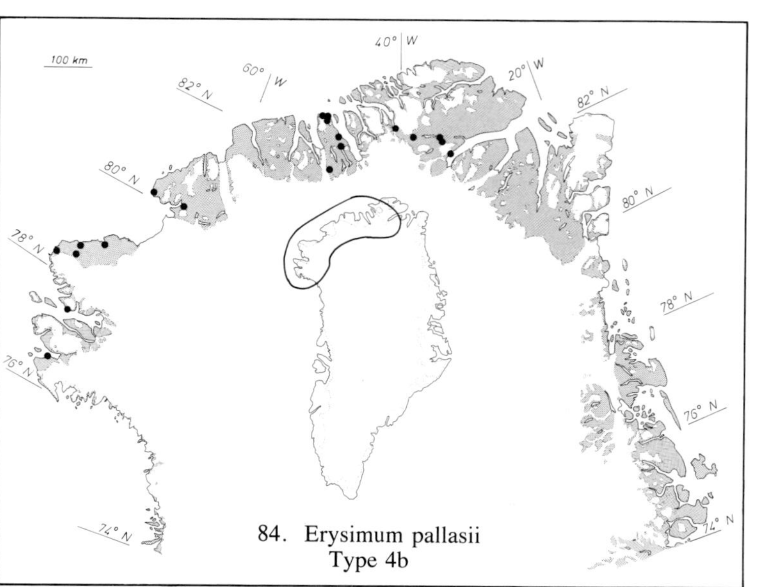

84. Erysimum pallasii
Type 4b

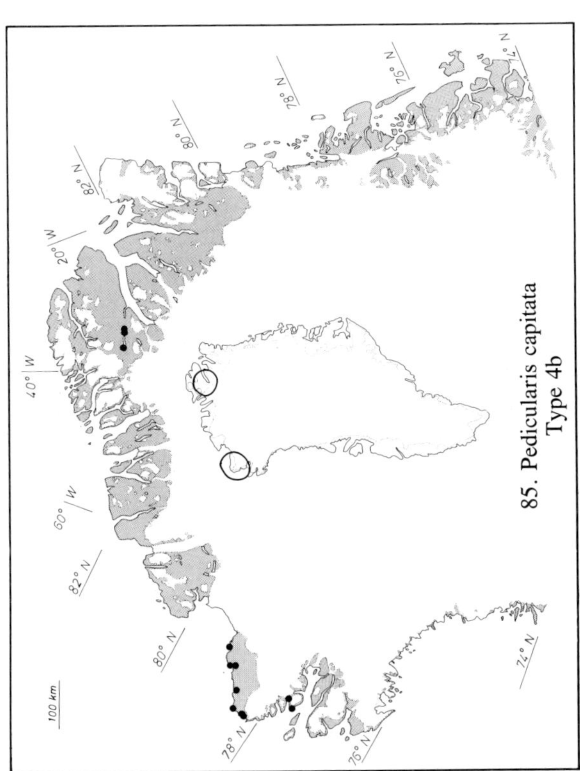

85. Pedicularis capitata
Type 4b

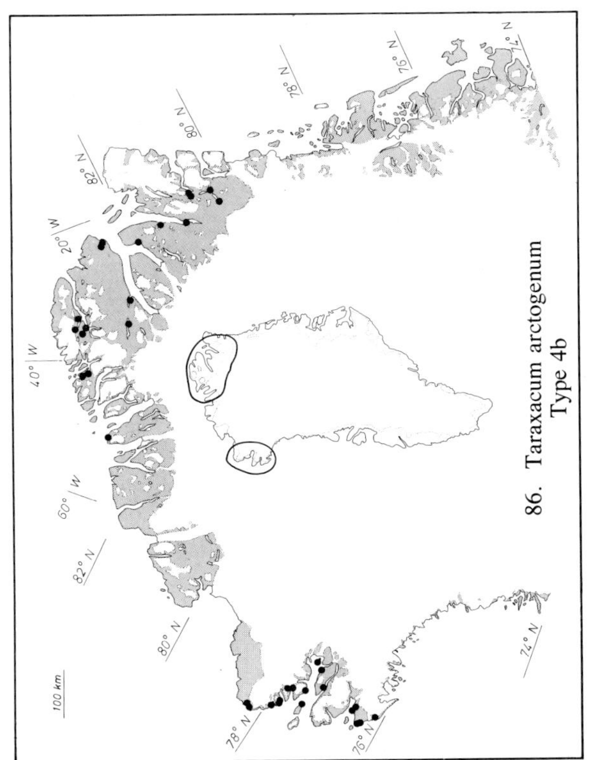

86. Taraxacum arctogenum
Type 4b

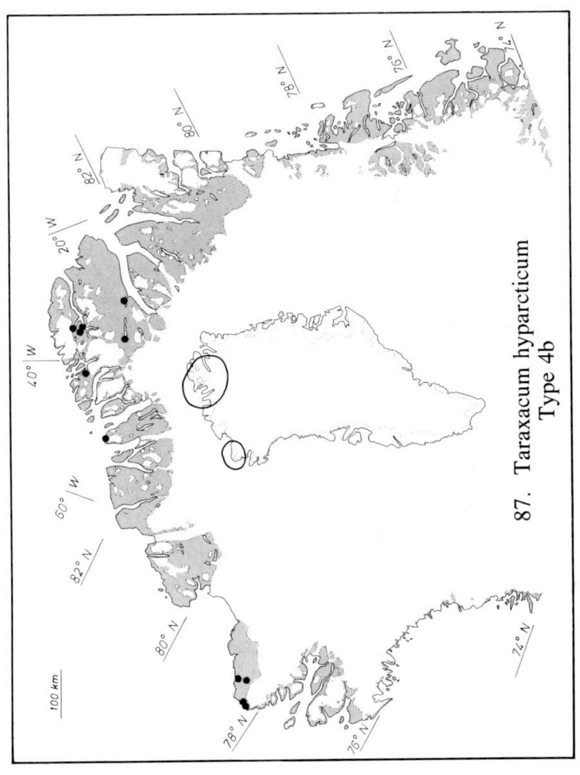

87. Taraxacum hyparcticum
Type 4b

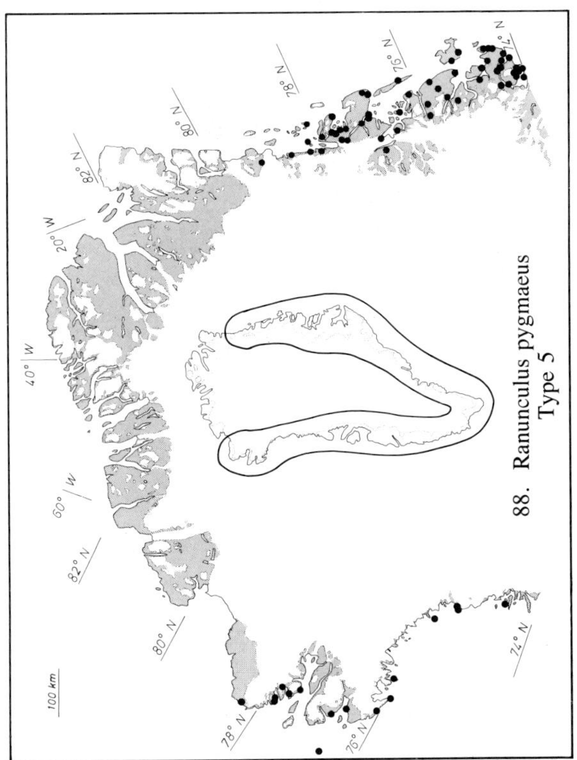

88. Ranunculus pygmaeus
Type 5

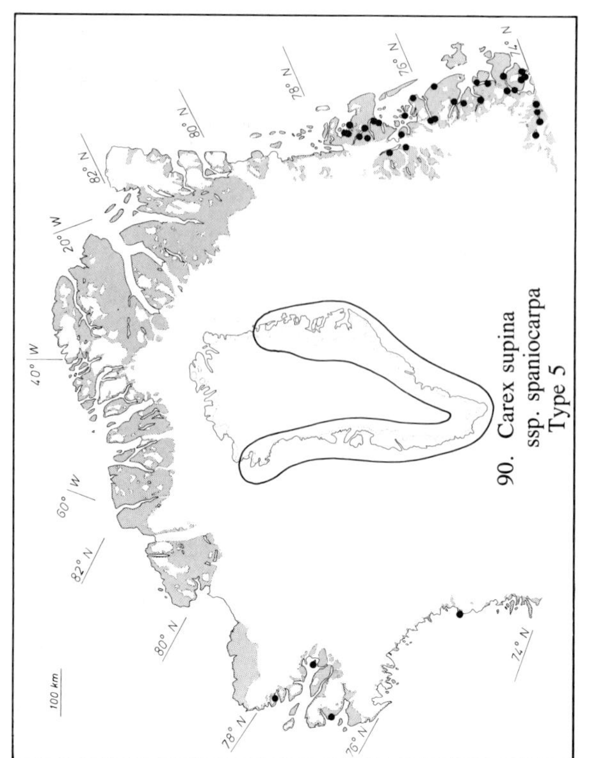

90. Carex supina ssp. spaniocarpa Type 5

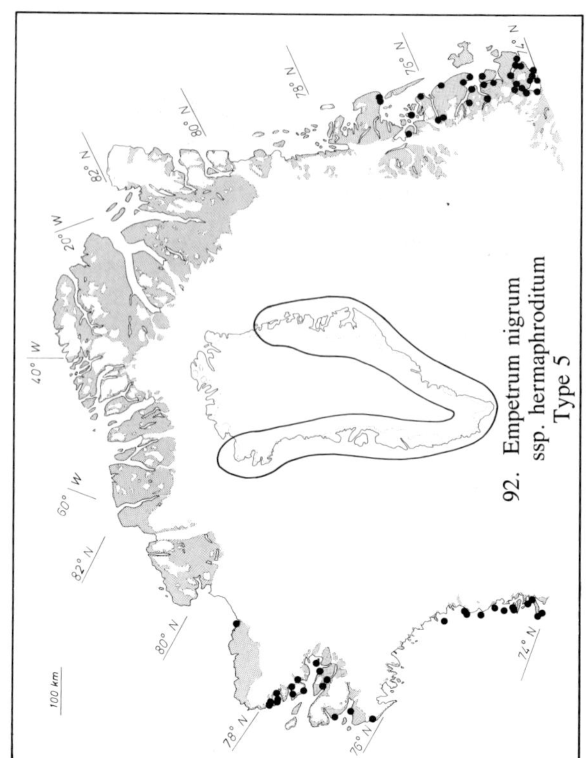

92. Empetrum nigrum ssp. hermaphroditum Type 5

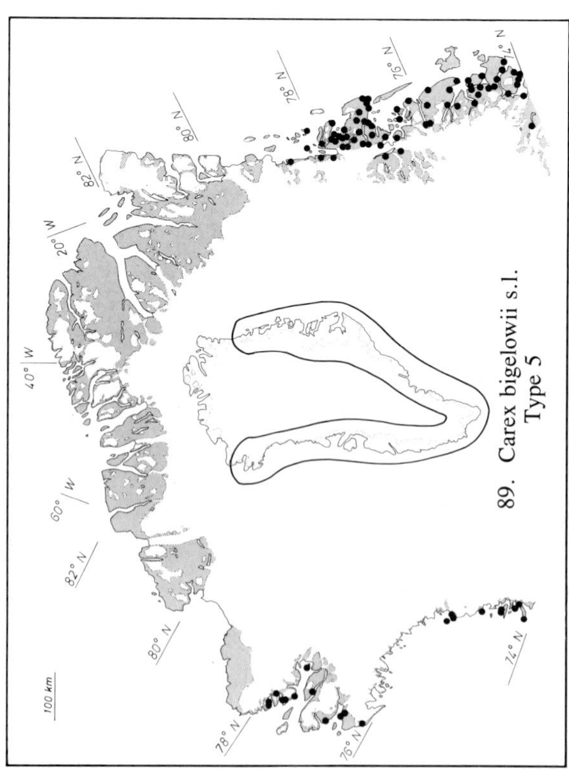

89. Carex bigelowii s.l. Type 5

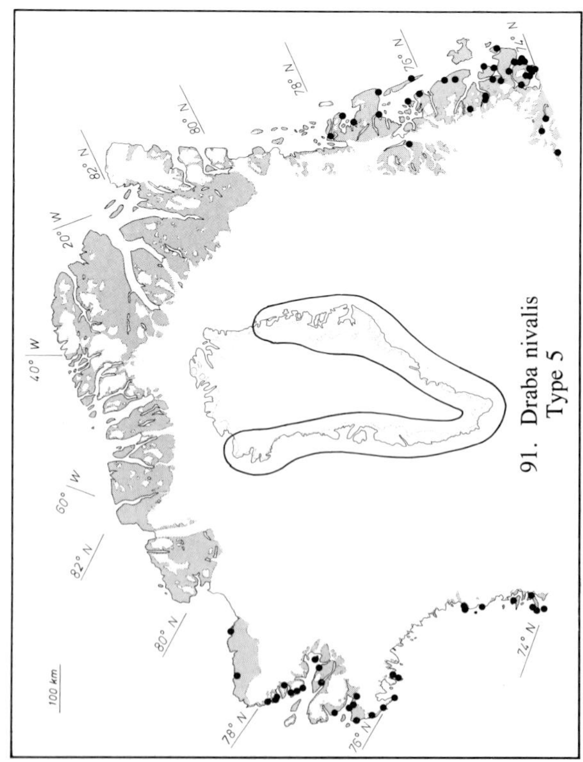

91. Draba nivalis Type 5

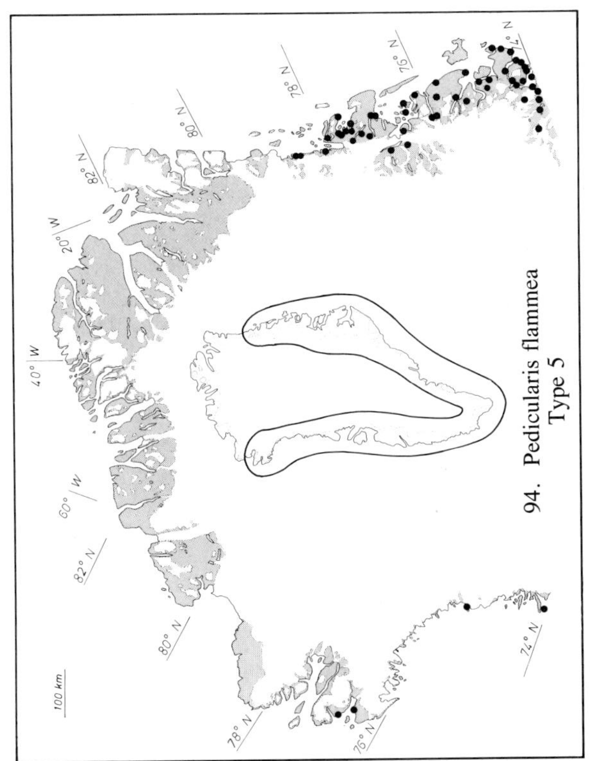

94. Pedicularis flammea
Type 5

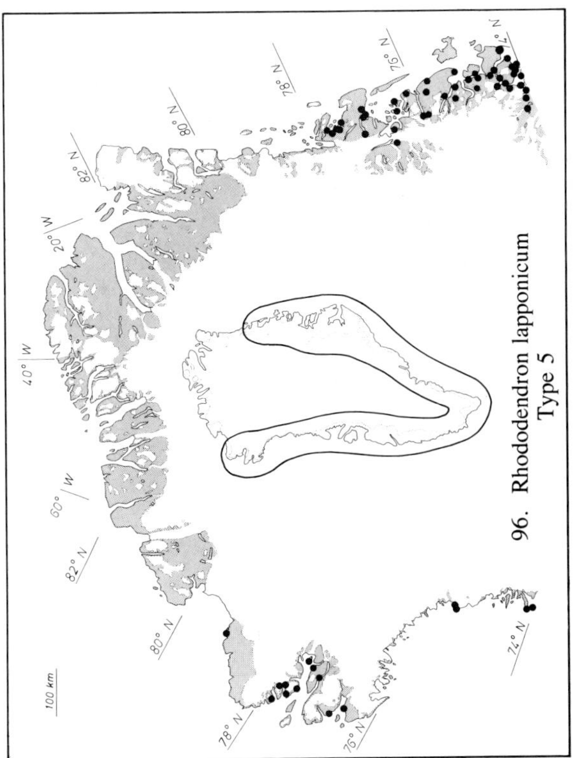

96. Rhododendron lapponicum
Type 5

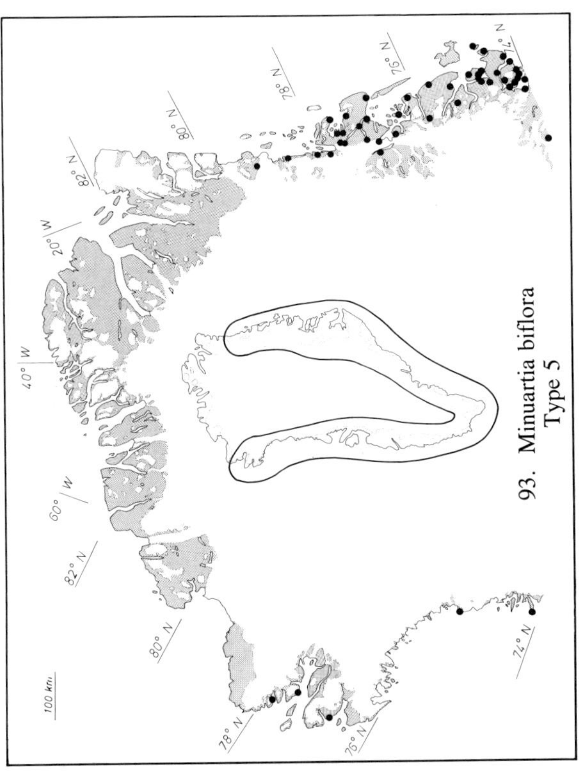

93. Minuartia biflora
Type 5

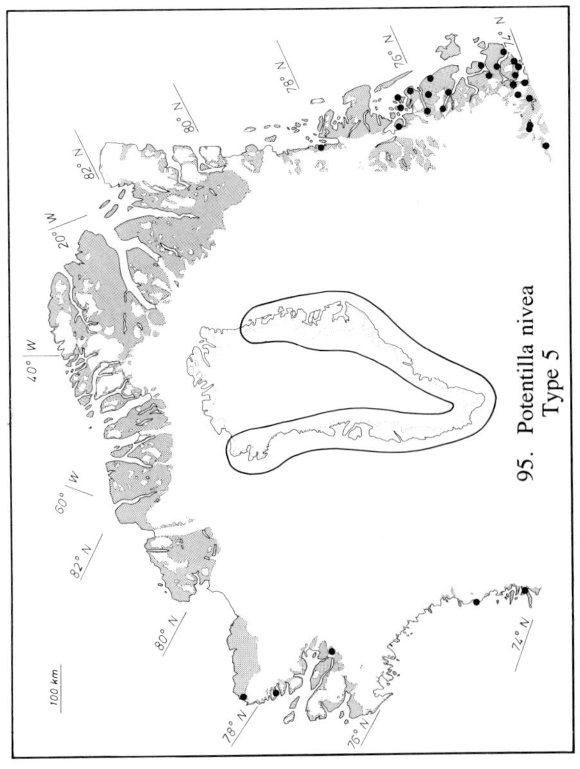

95. Potentilla nivea
Type 5

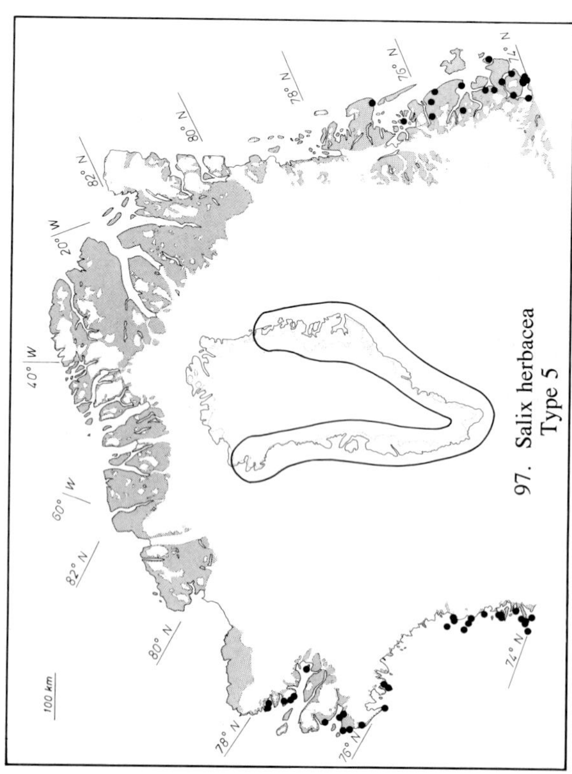

97. Salix herbacea
Type 5

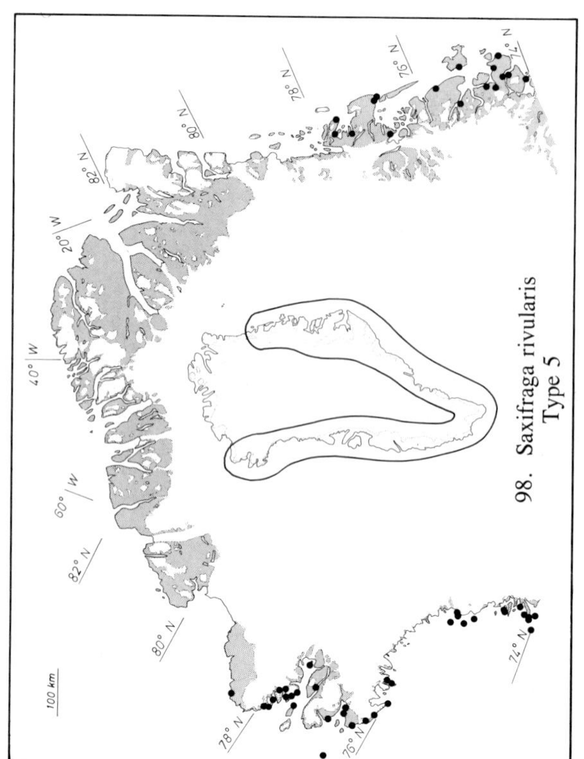

98. Saxifraga rivularis
Type 5

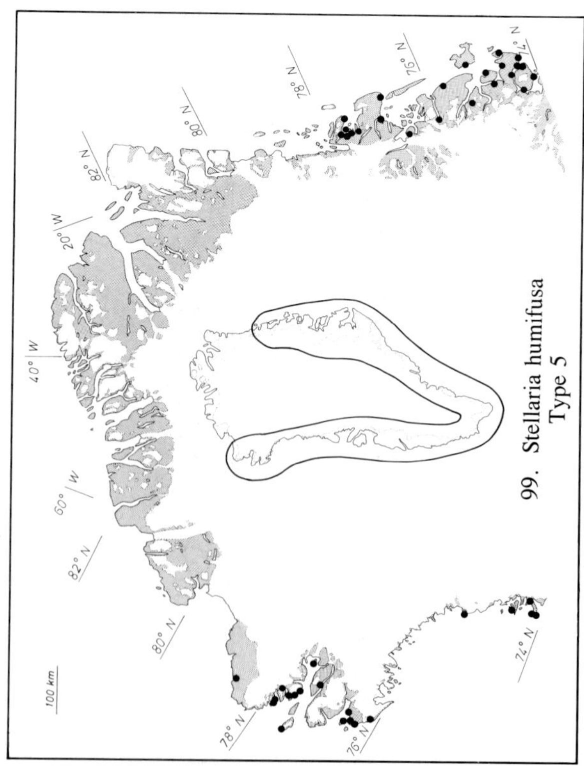

99. Stellaria humifusa
Type 5

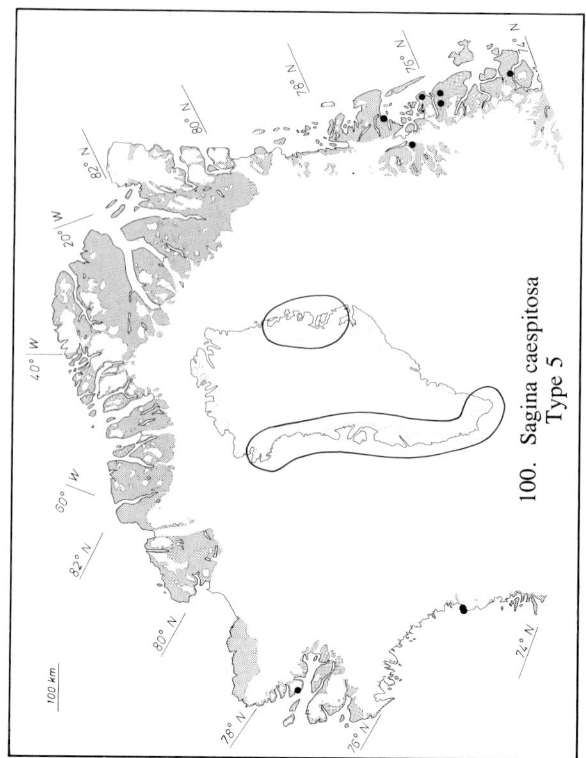

100. Sagina caespitosa
Type 5

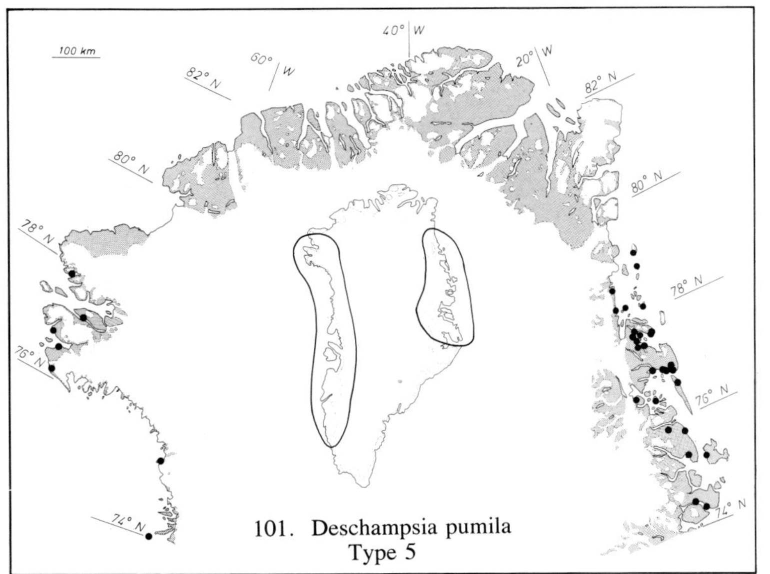

101. Deschampsia pumila
Type 5

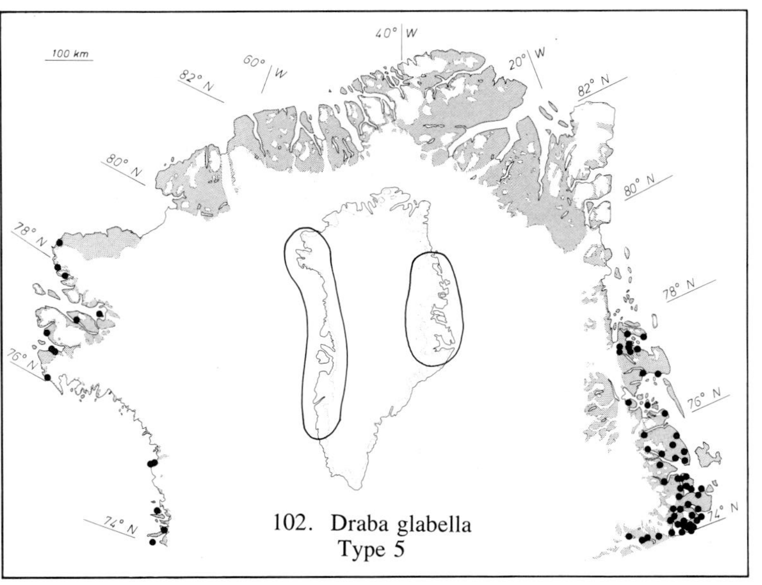

102. Draba glabella
Type 5

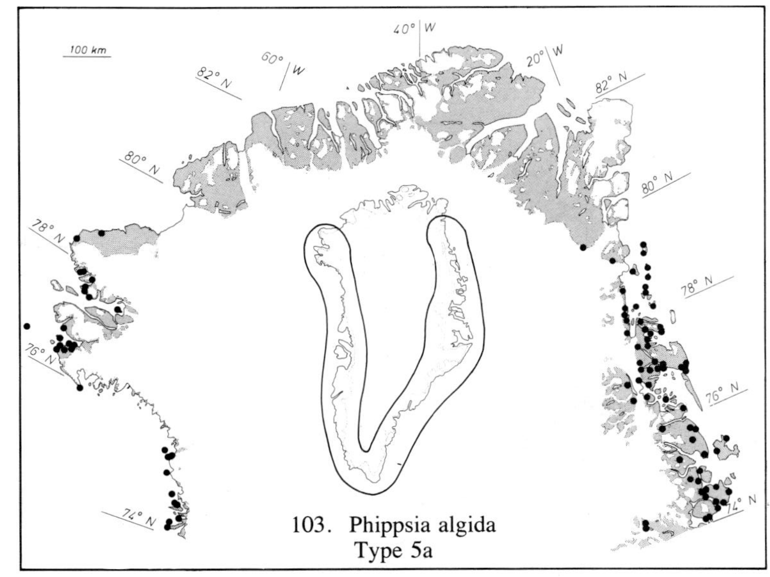

103. Phippsia algida
Type 5a

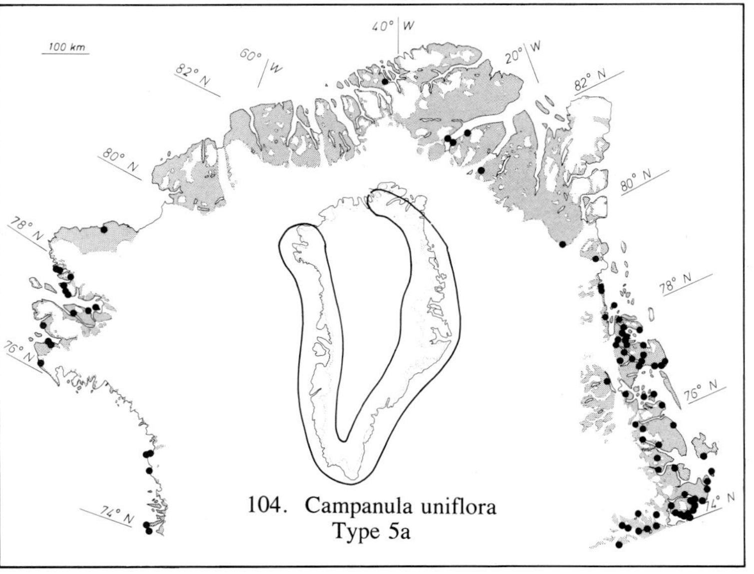

104. Campanula uniflora
Type 5a

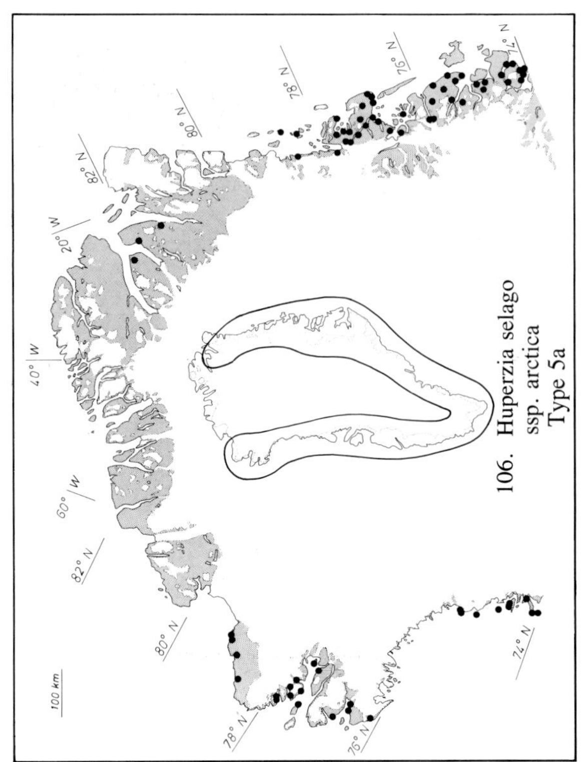

106. Huperzia selago ssp. arctica Type 5a

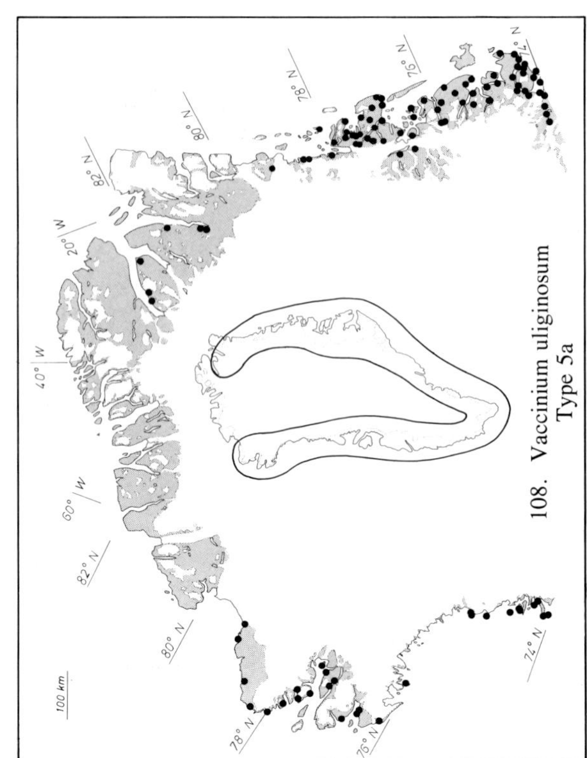

108. Vaccinium uliginosum Type 5a

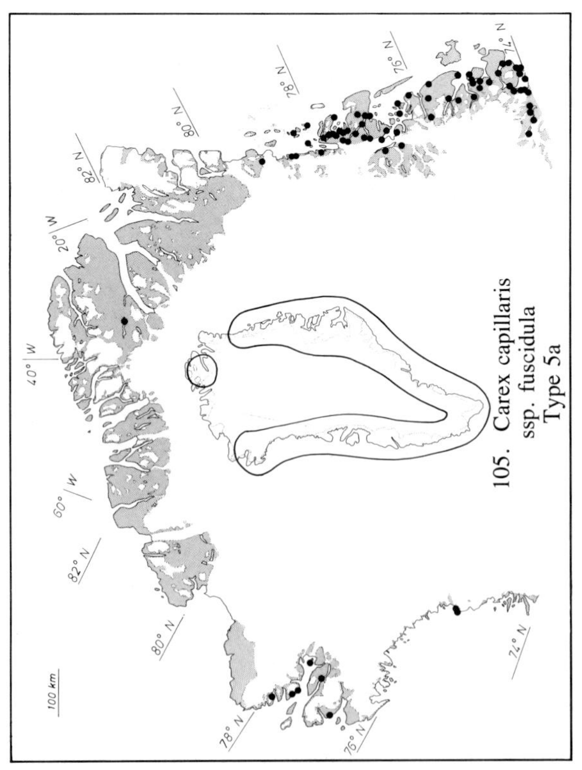

105. Carex capillaris ssp. fuscidula Type 5a

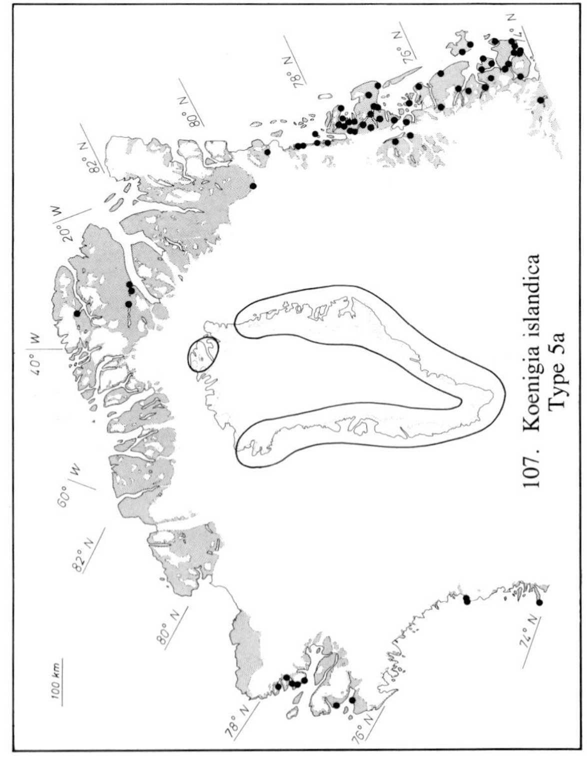

107. Koenigia islandica Type 5a

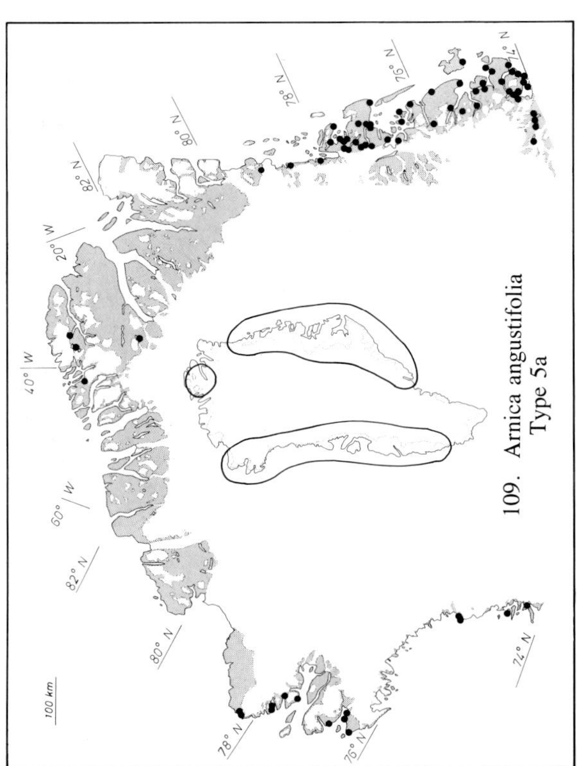

109. Arnica angustifolia
Type 5a

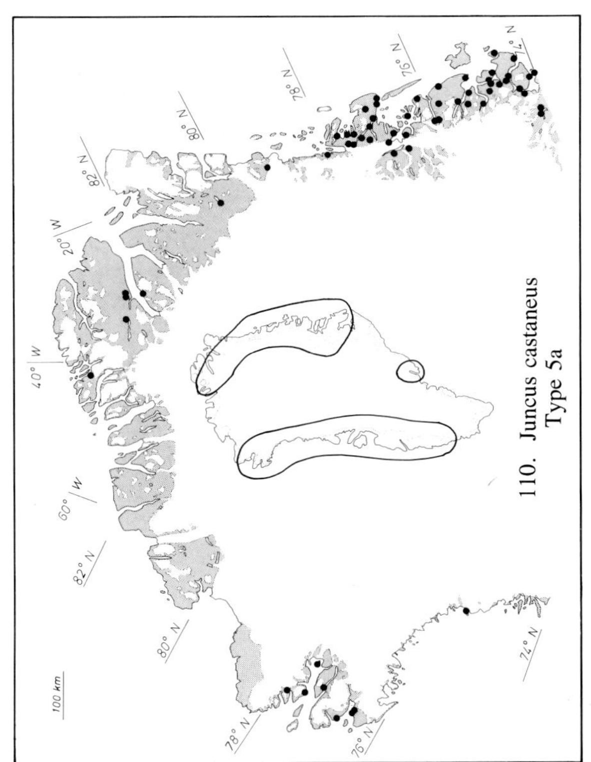

110. Juncus castaneus
Type 5a

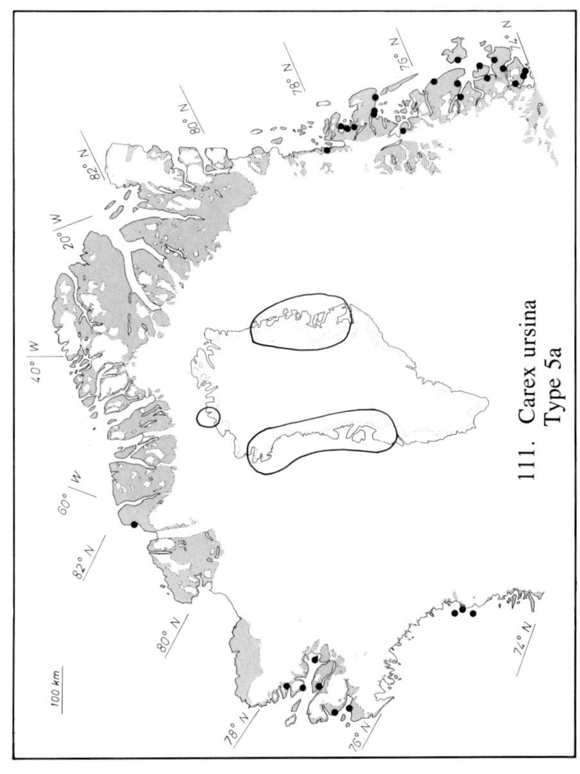

111. Carex ursina
Type 5a

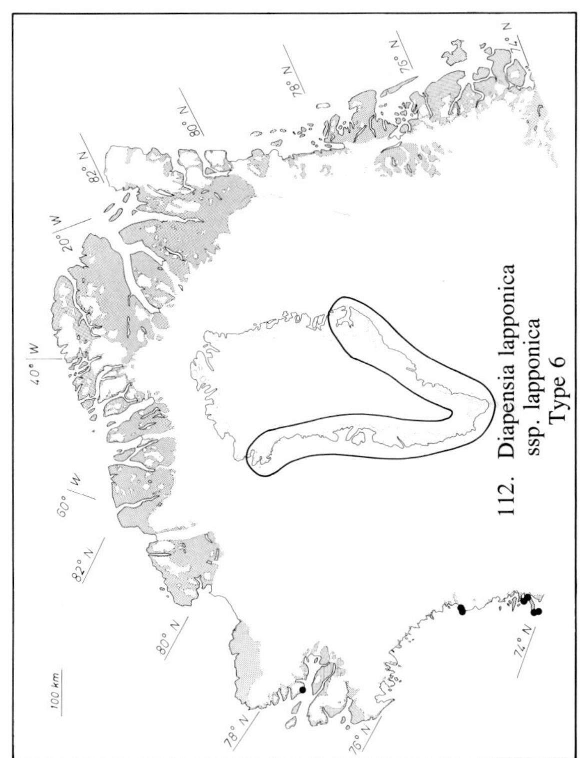

112. Diapensia lapponica
ssp. lapponica
Type 6

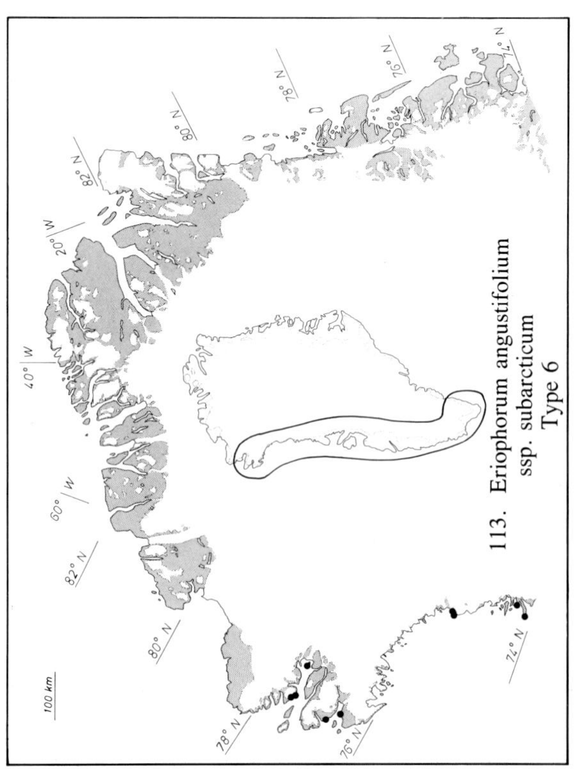

113. Eriophorum angustifolium ssp. subarcticum
Type 6

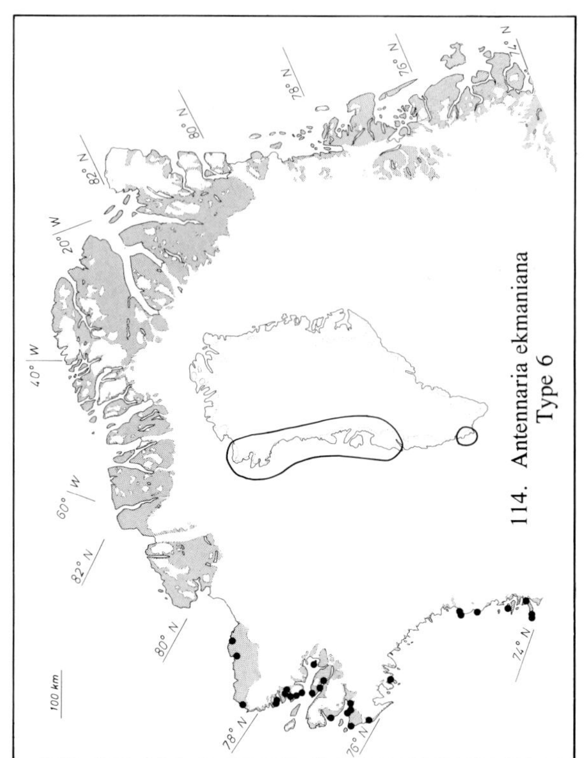

114. Antennaria ekmaniana
Type 6

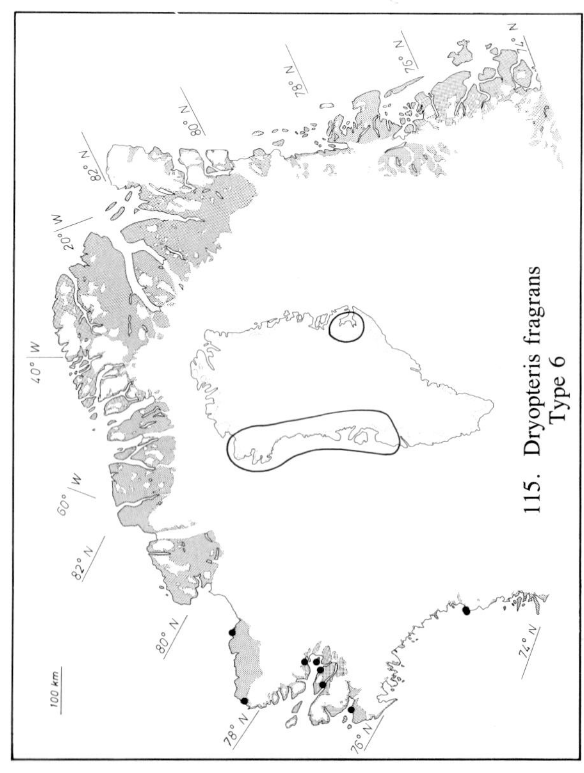

115. Dryopteris fragrans
Type 6

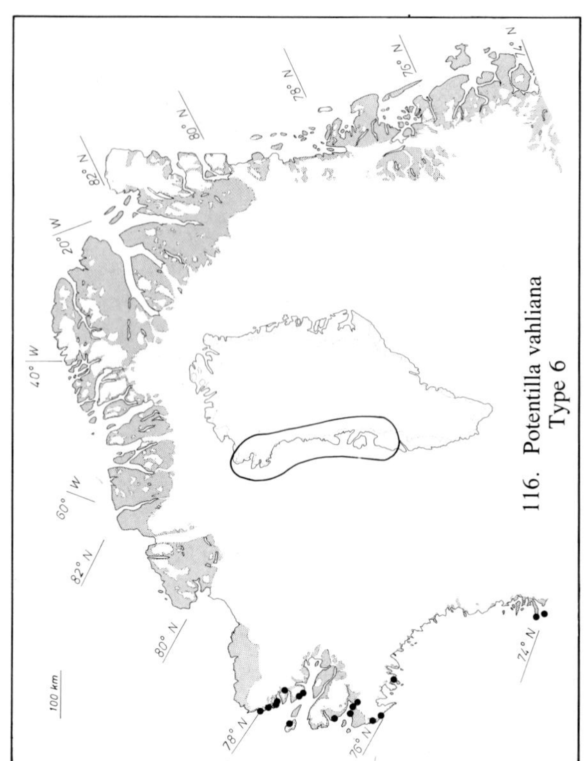

116. Potentilla vahliana
Type 6

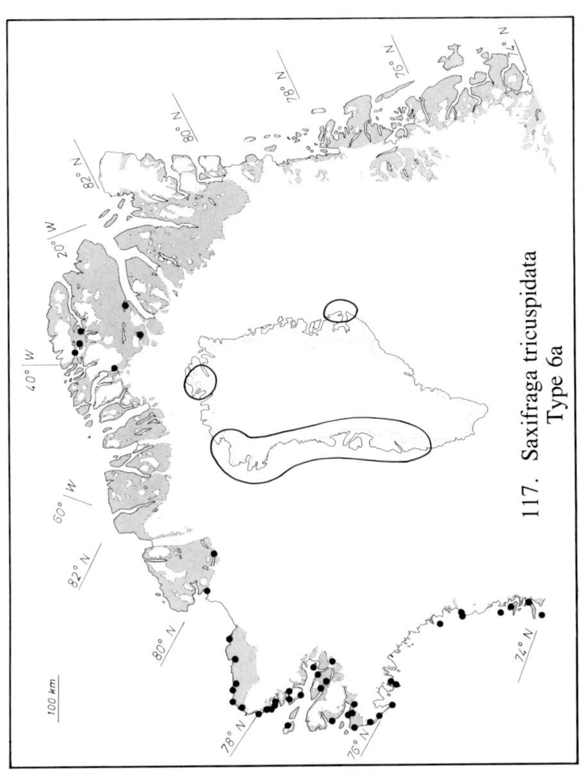

117. Saxifraga tricuspidata
Type 6a

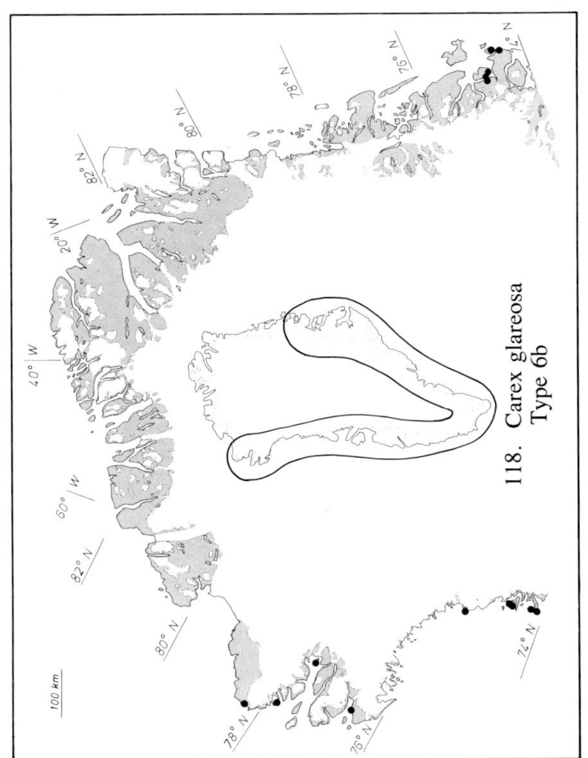

118. Carex glareosa
Type 6b

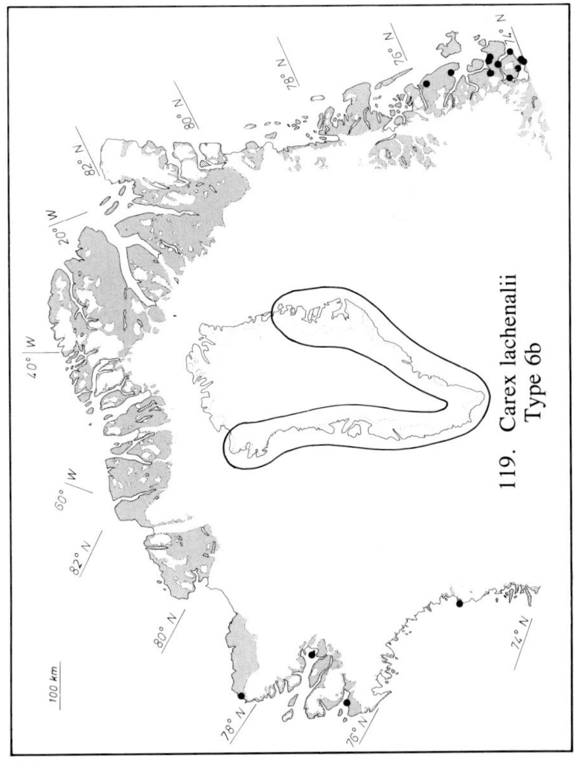

119. Carex lachenalii
Type 6b

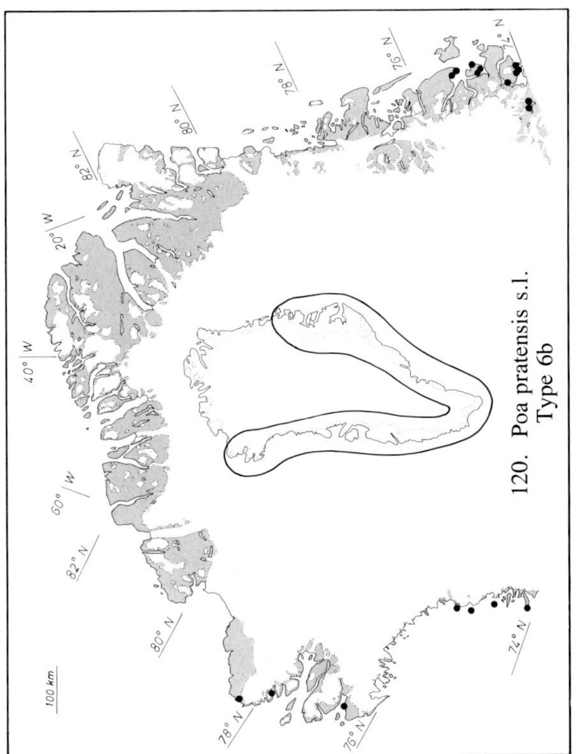

120. Poa pratensis s.l.
Type 6b

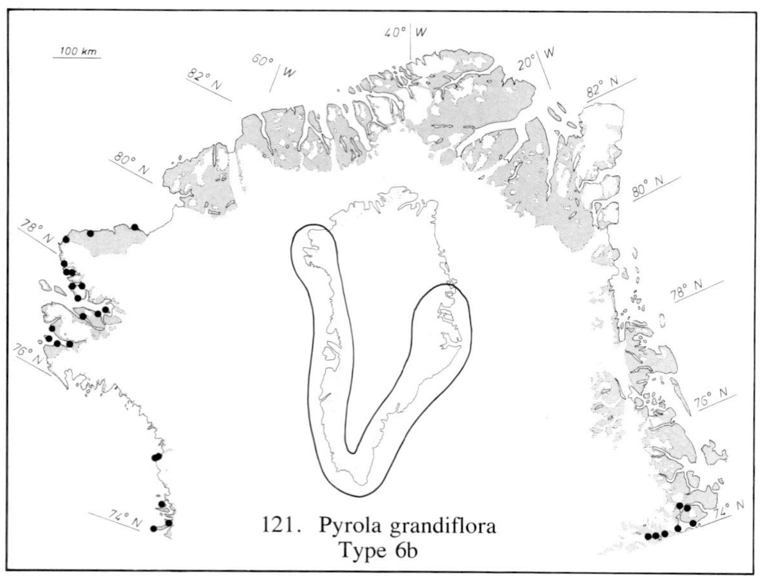

121. Pyrola grandiflora
Type 6b

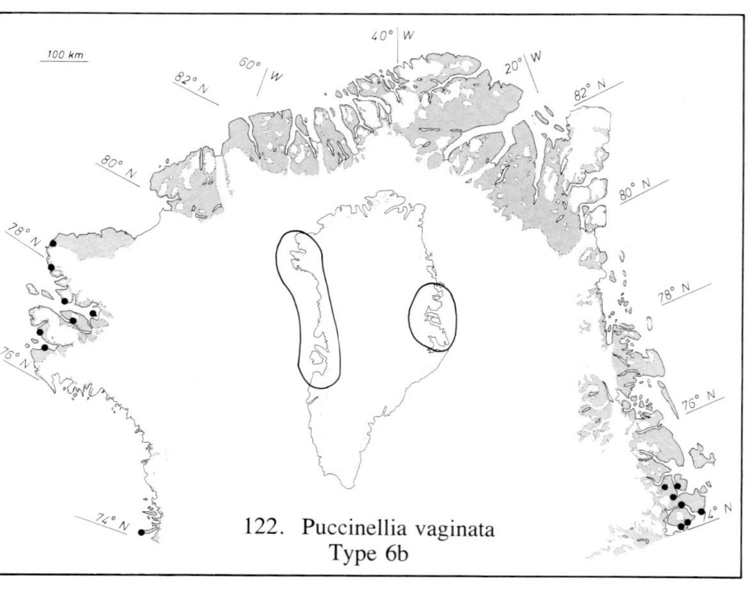

122. Puccinellia vaginata
Type 6b

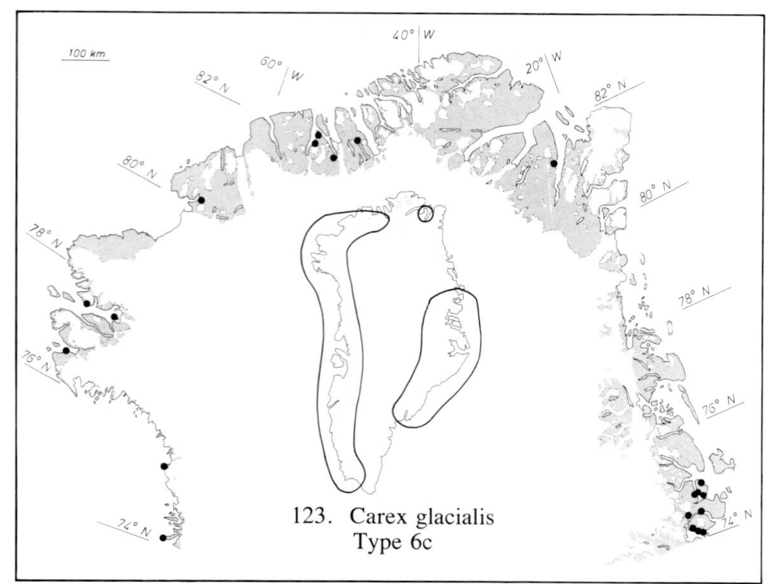

123. Carex glacialis
Type 6c

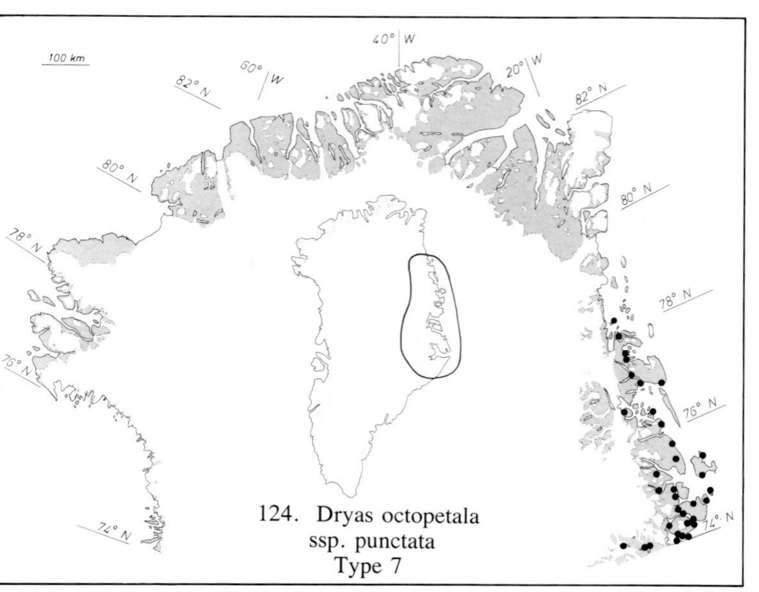

124. Dryas octopetala
ssp. punctata
Type 7

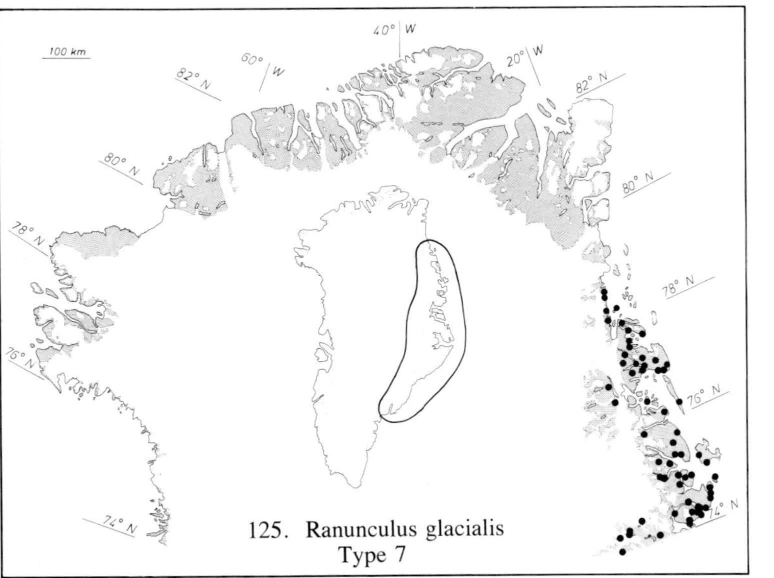

125. Ranunculus glacialis
Type 7

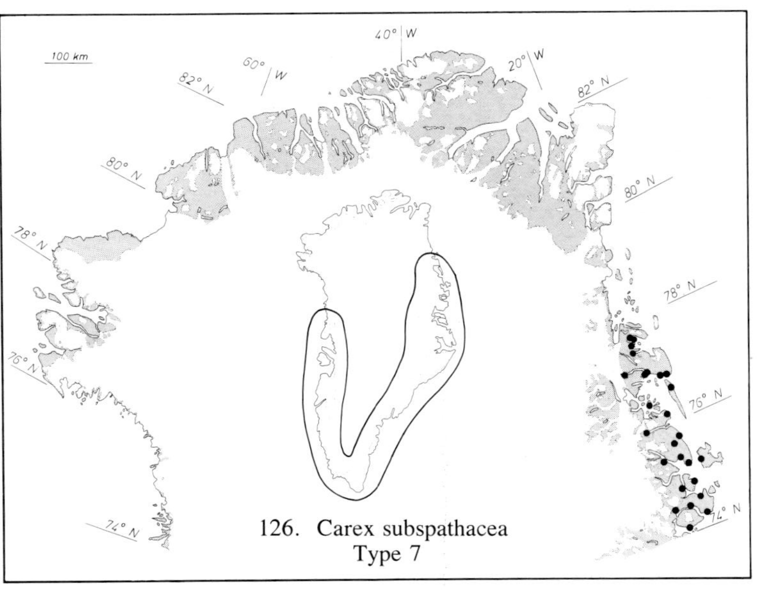

126. Carex subspathacea
Type 7

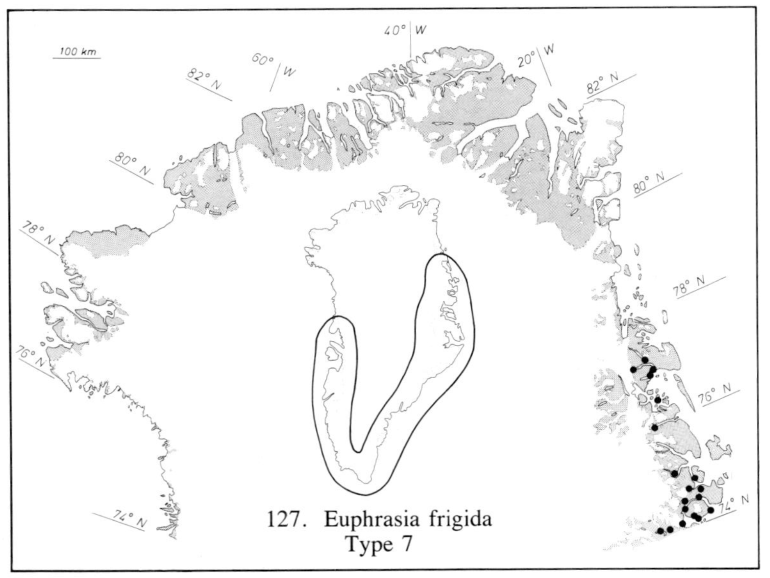

127. Euphrasia frigida
Type 7

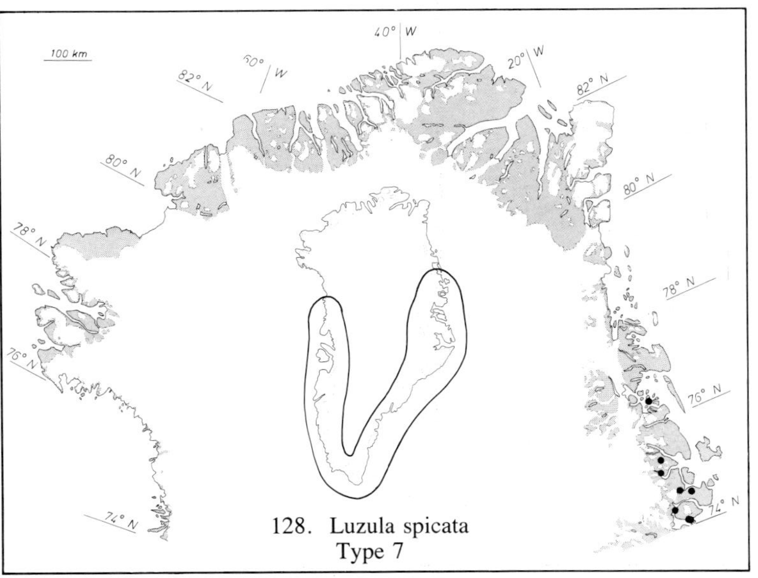

128. Luzula spicata
Type 7

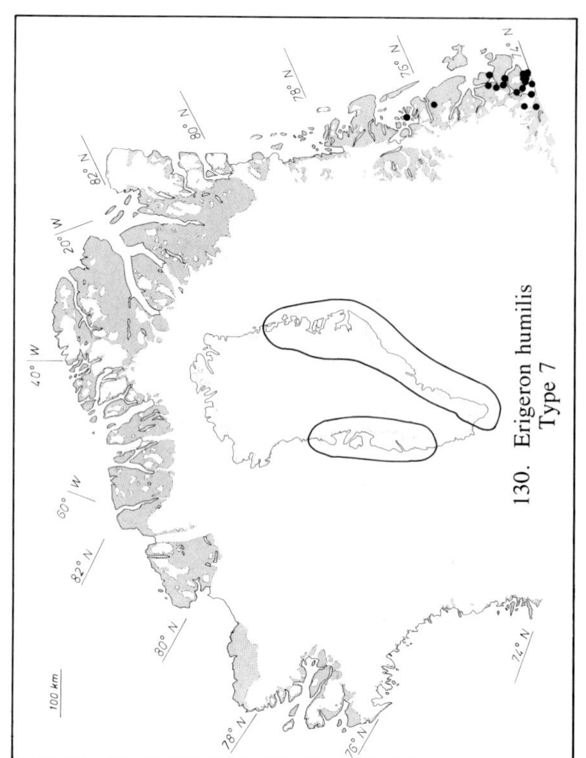
130. Erigeron humilis
Type 7

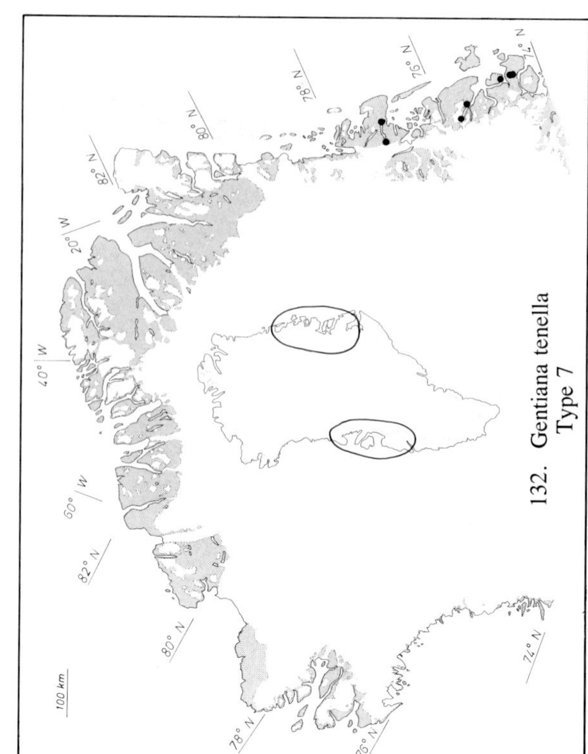
132. Gentiana tenella
Type 7

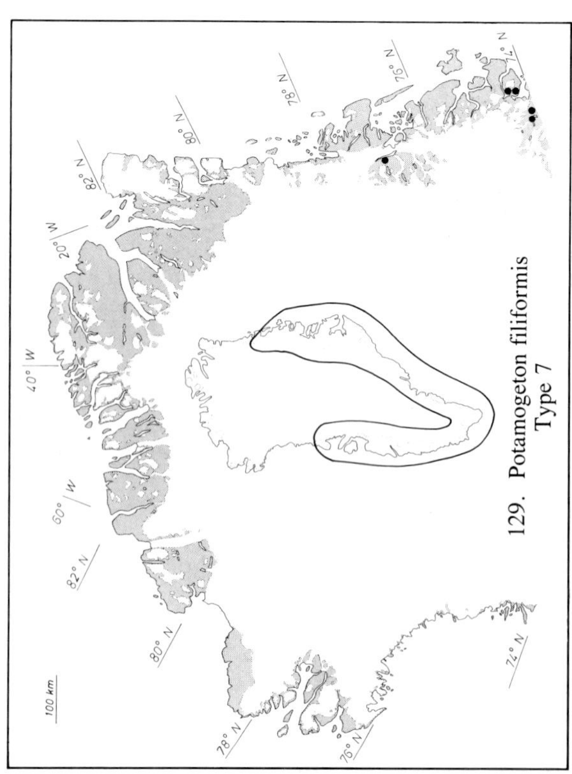
129. Potamogeton filiformis
Type 7

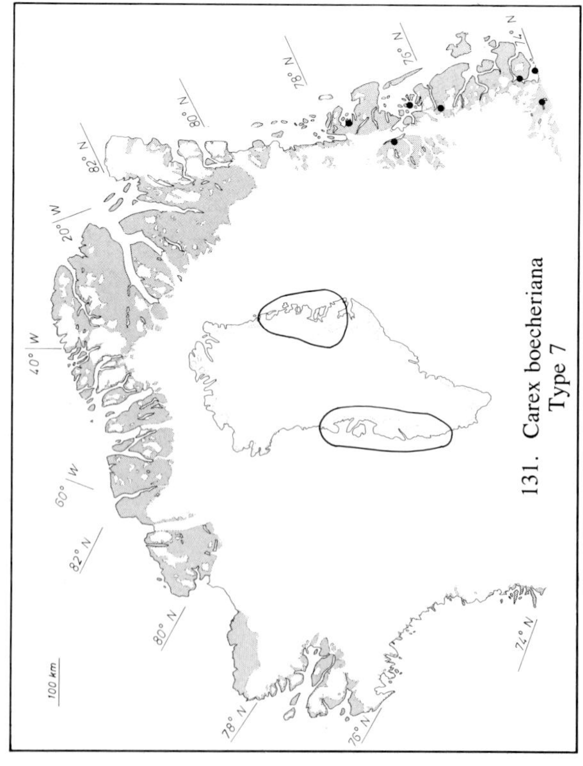
131. Carex boecheriana
Type 7

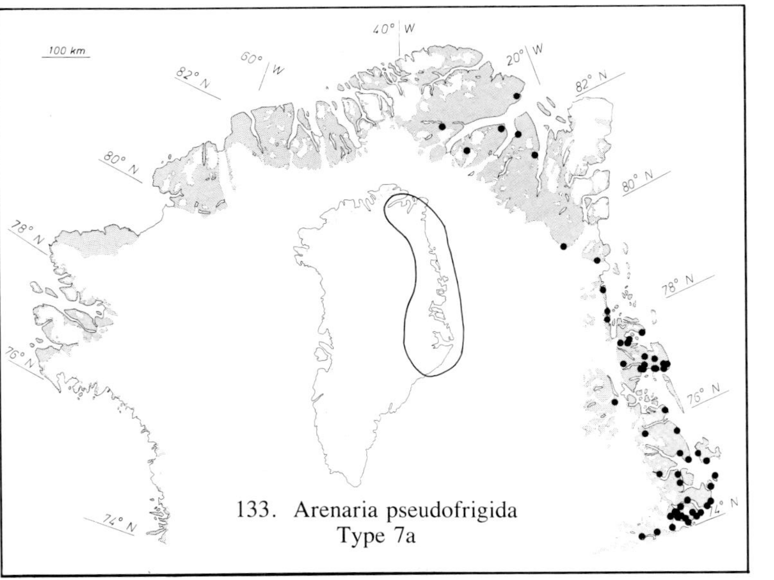

133. Arenaria pseudofrigida
Type 7a

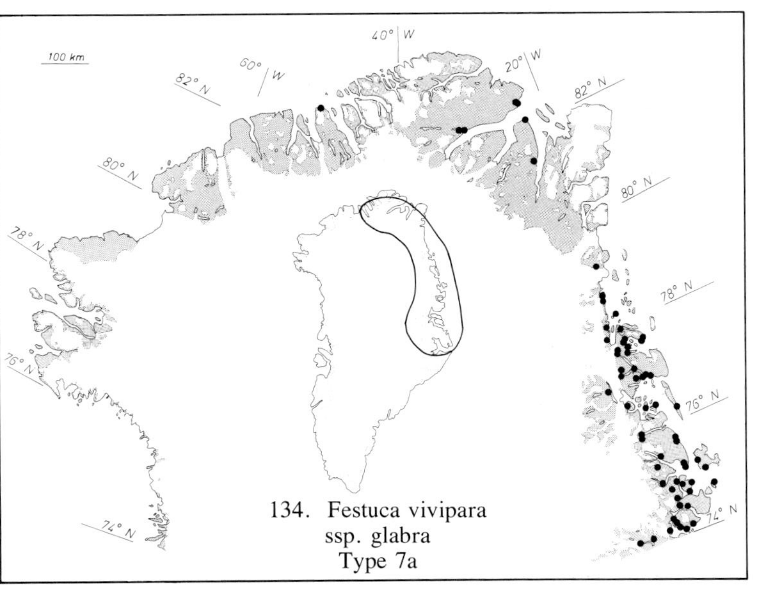

134. Festuca vivipara
ssp. glabra
Type 7a

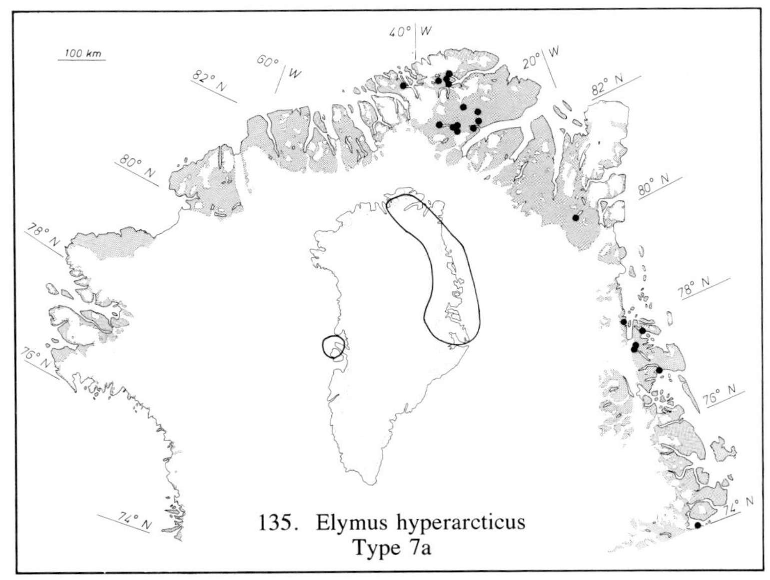

135. Elymus hyperarcticus
Type 7a

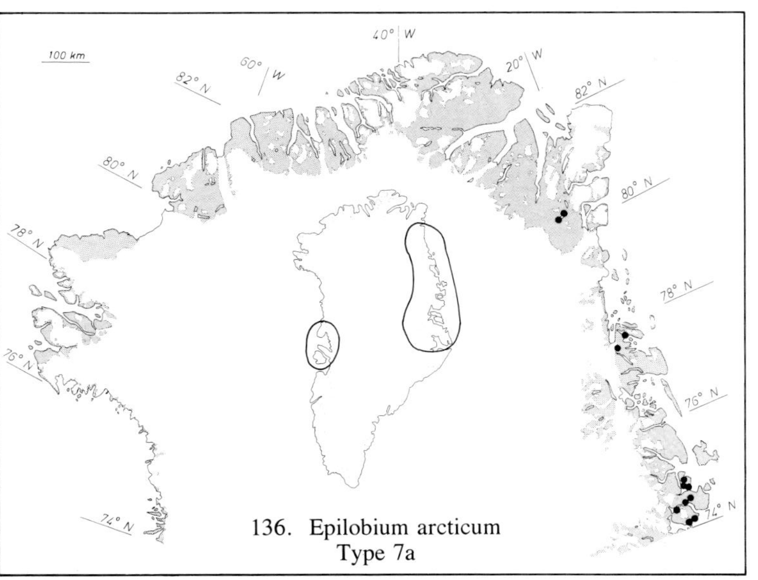

136. Epilobium arcticum
Type 7a

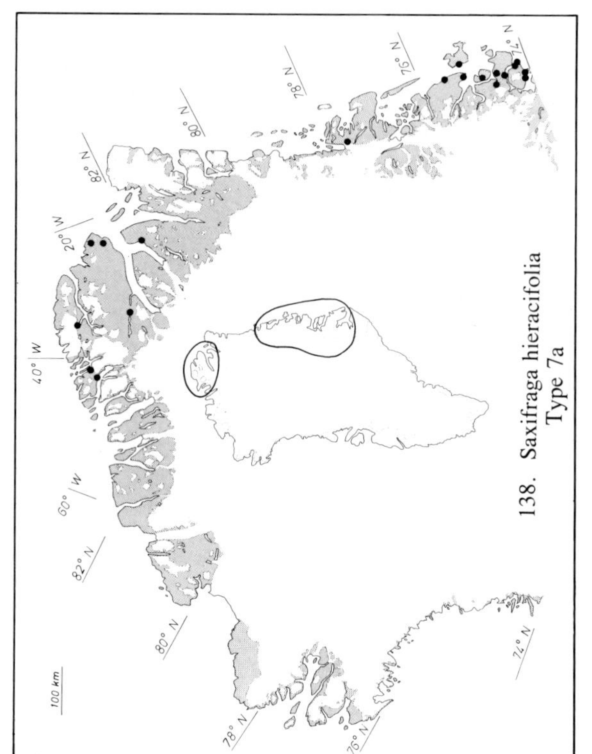

138. Saxifraga hieracifolia
Type 7a

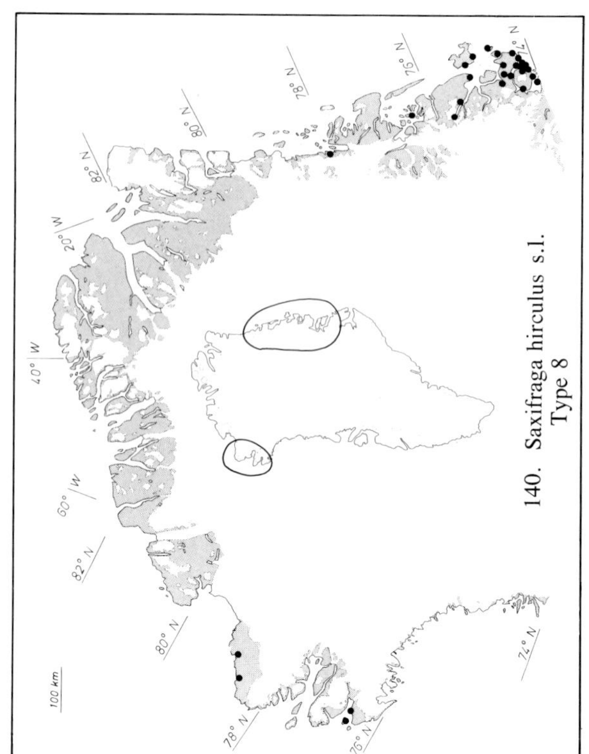

140. Saxifraga hirculus s.l.
Type 8

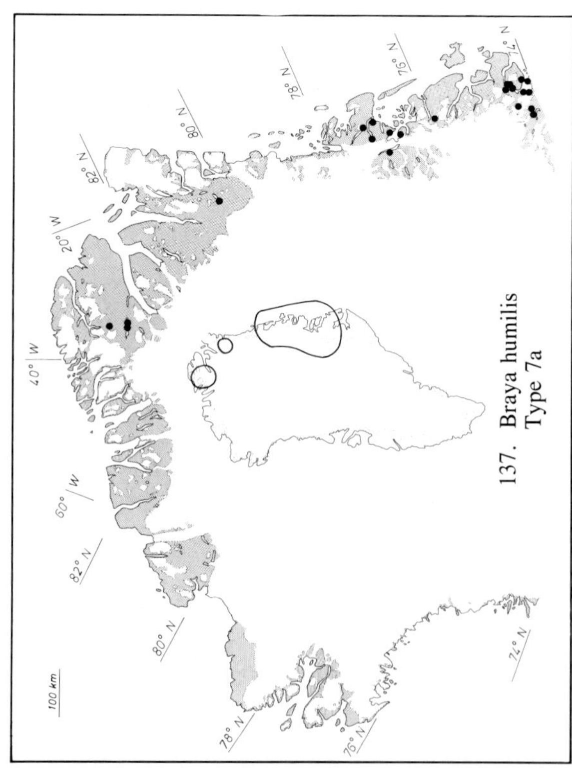

137. Braya humilis
Type 7a

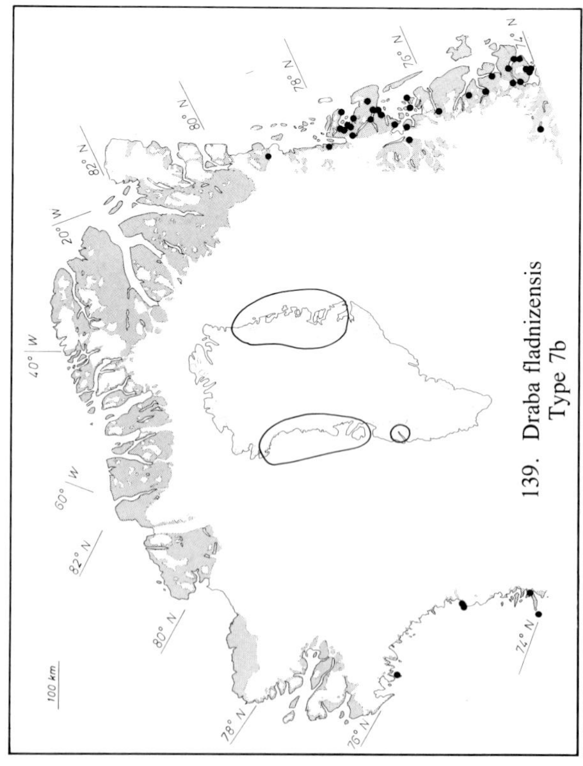

139. Draba fladnizensis
Type 7b

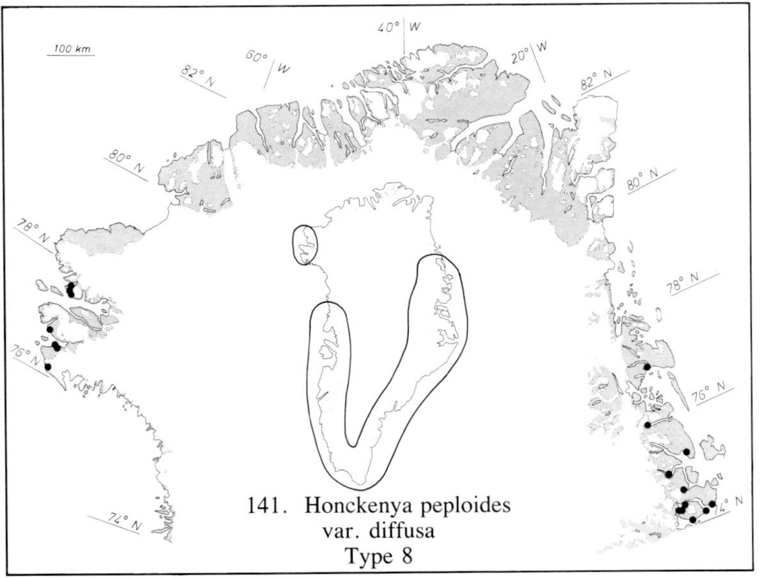

141. Honckenya peploides
var. diffusa
Type 8

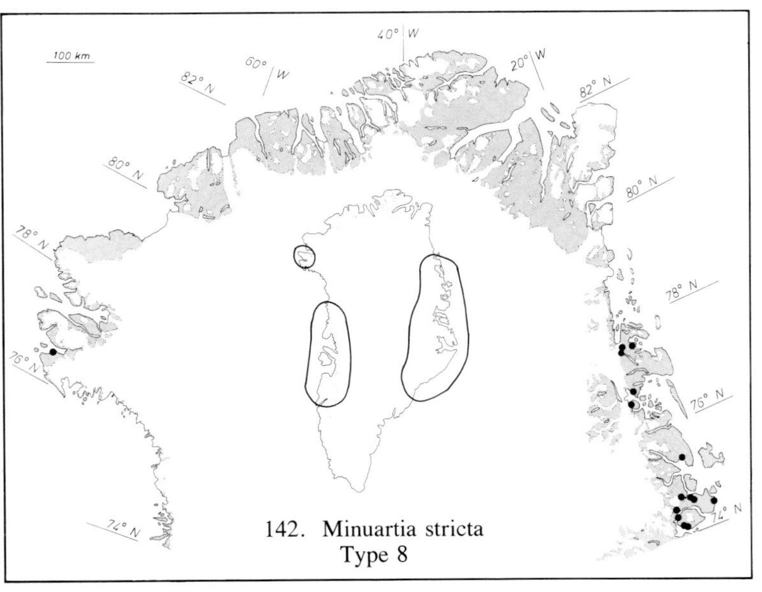

142. Minuartia stricta
Type 8

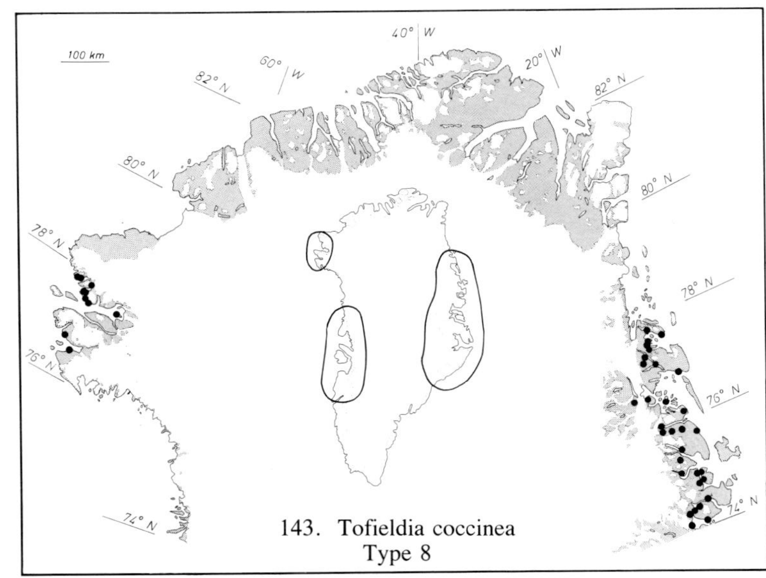

143. Tofieldia coccinea
Type 8

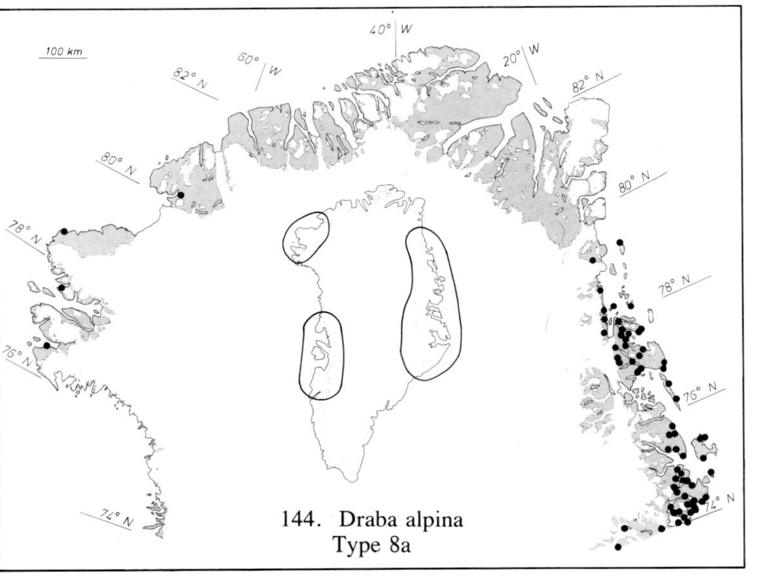

144. Draba alpina
Type 8a

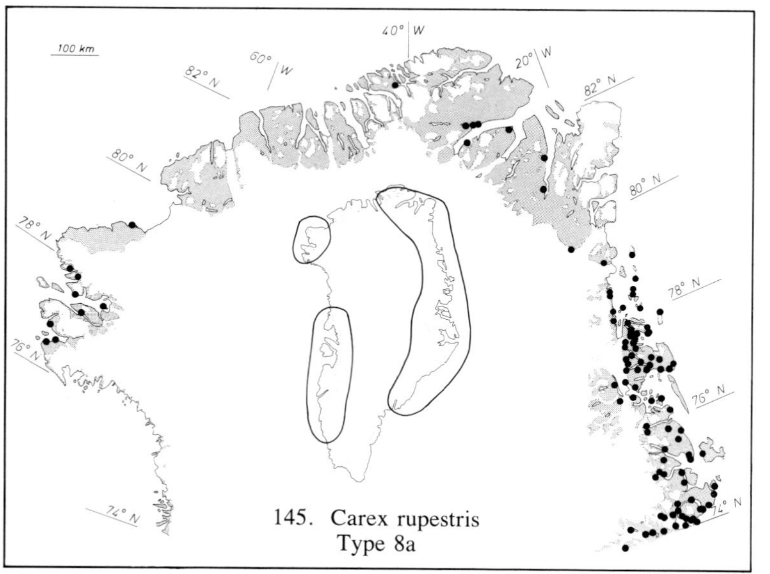
145. Carex rupestris
Type 8a

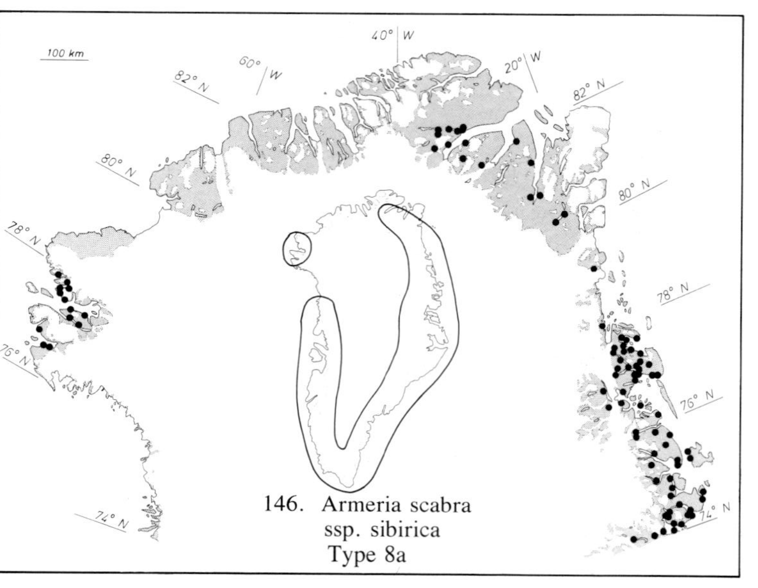
146. Armeria scabra
ssp. sibirica
Type 8a

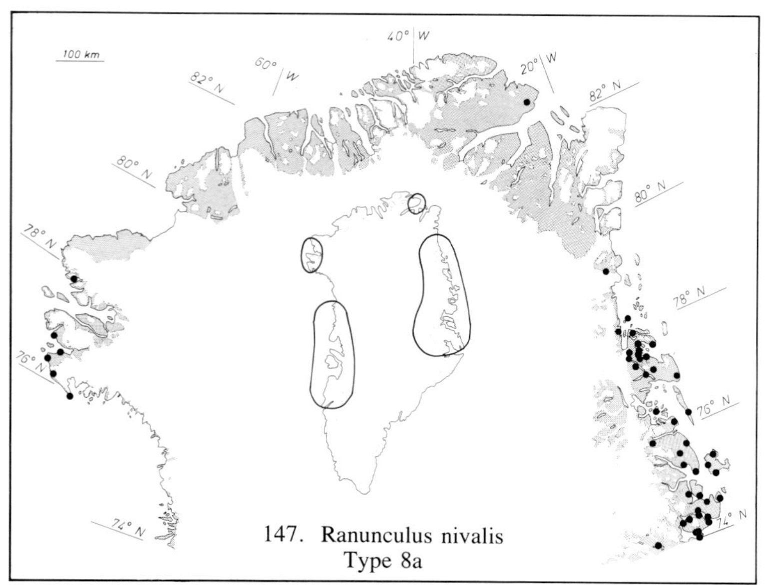
147. Ranunculus nivalis
Type 8a

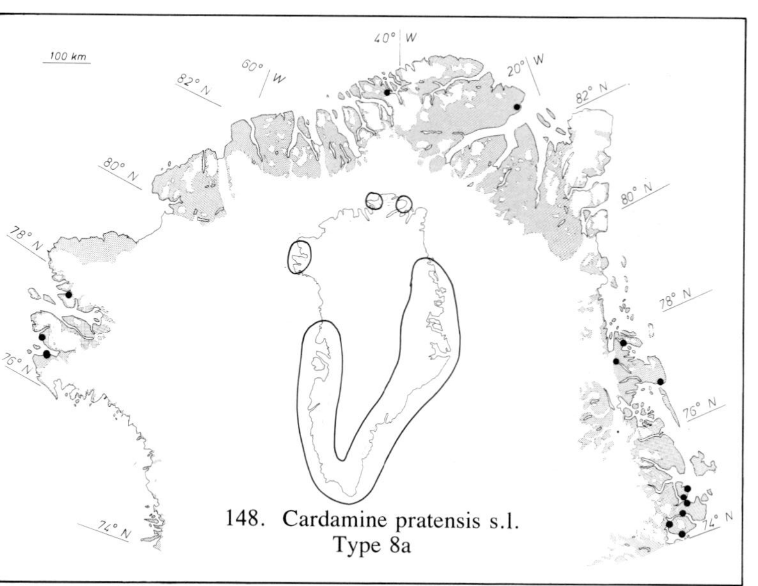
148. Cardamine pratensis s.l.
Type 8a

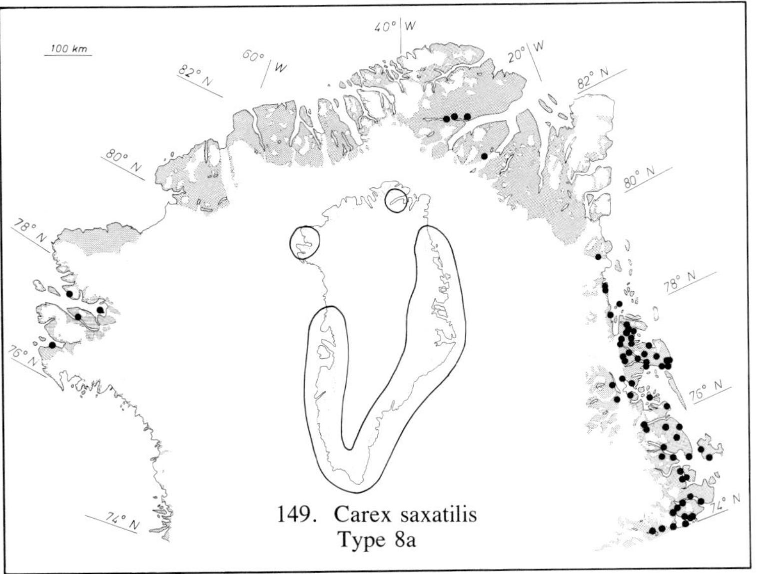

149. Carex saxatilis
Type 8a

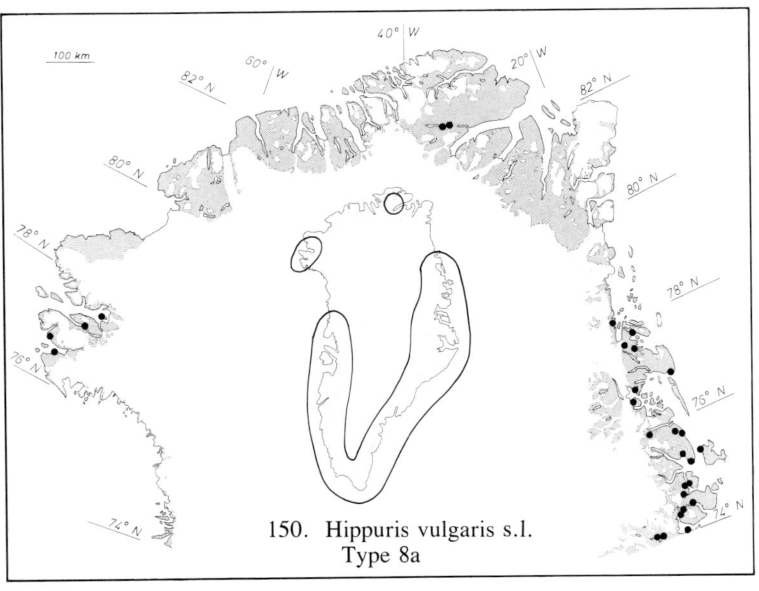

150. Hippuris vulgaris s.l.
Type 8a

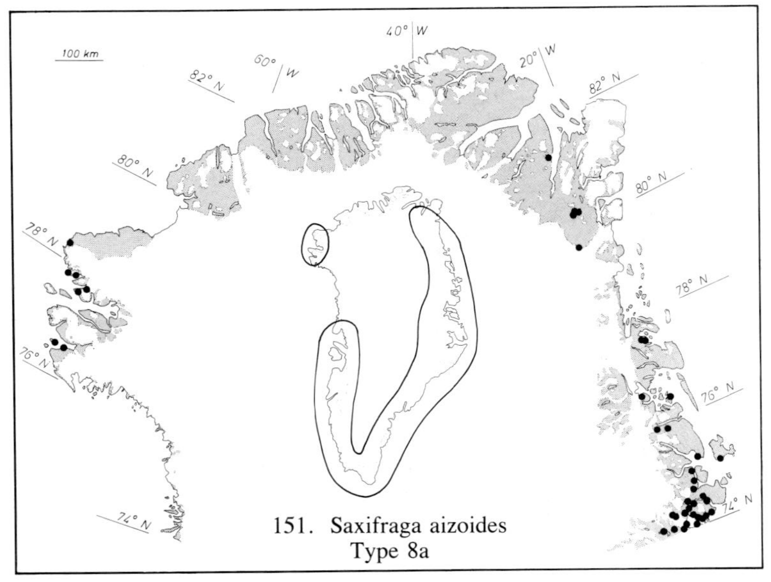

151. Saxifraga aizoides
Type 8a

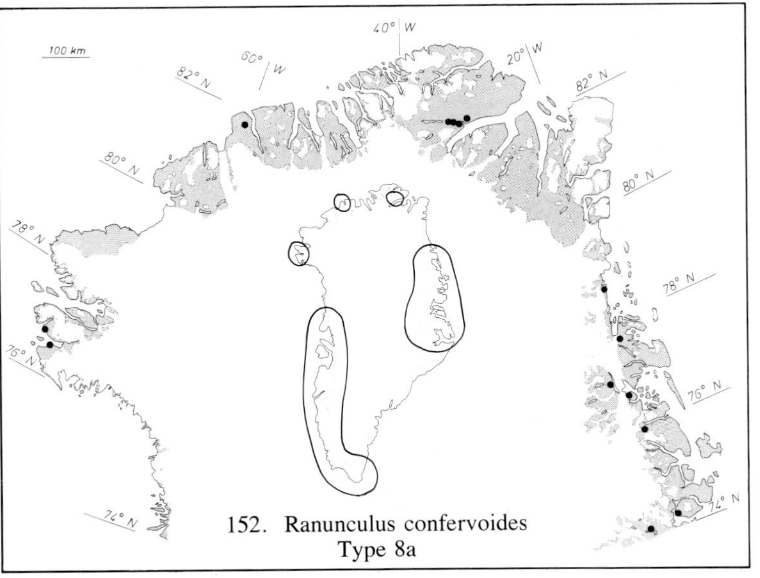

152. Ranunculus confervoides
Type 8a

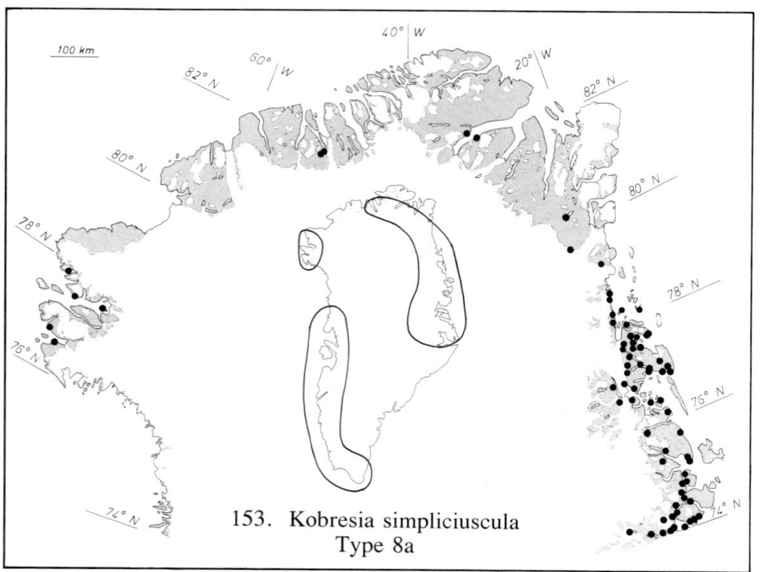

153. Kobresia simpliciuscula
Type 8a

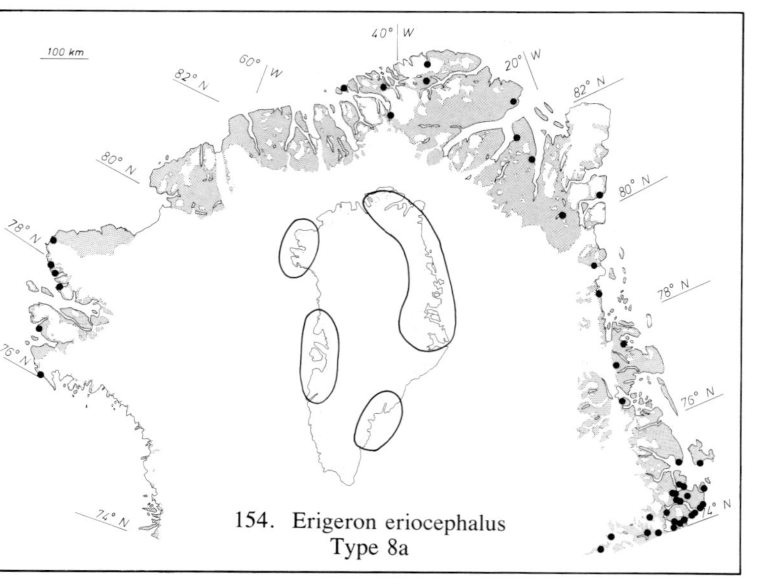

154. Erigeron eriocephalus
Type 8a

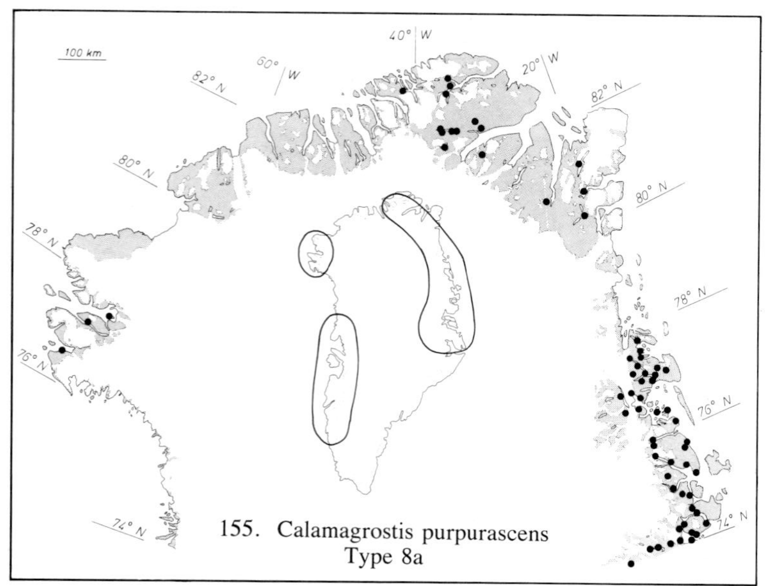

155. Calamagrostis purpurascens
Type 8a

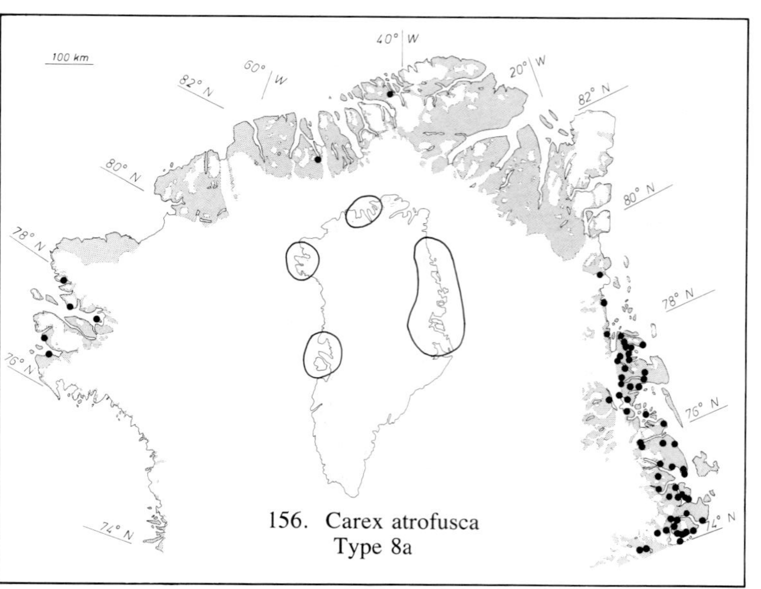

156. Carex atrofusca
Type 8a

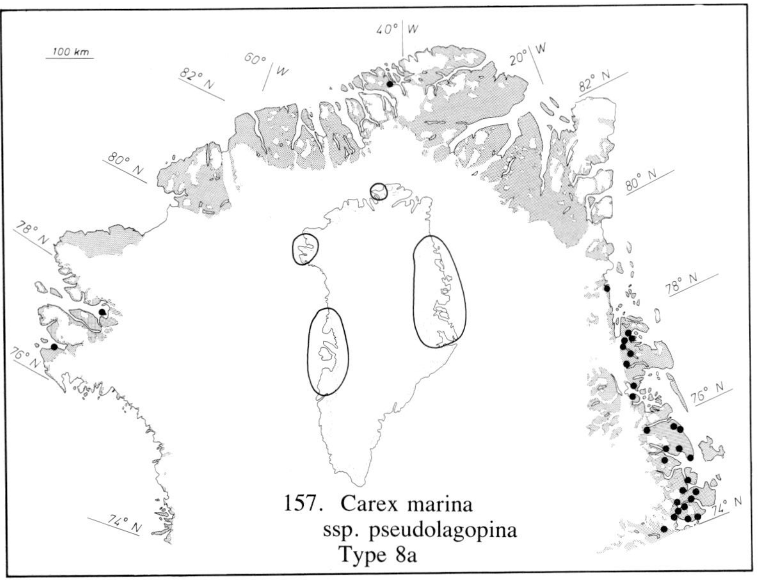

157. Carex marina
ssp. pseudolagopina
Type 8a

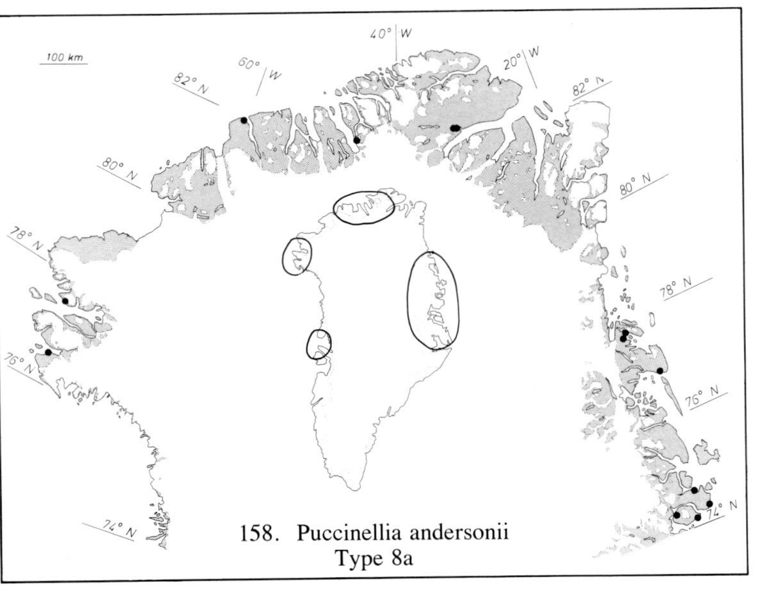

158. Puccinellia andersonii
Type 8a

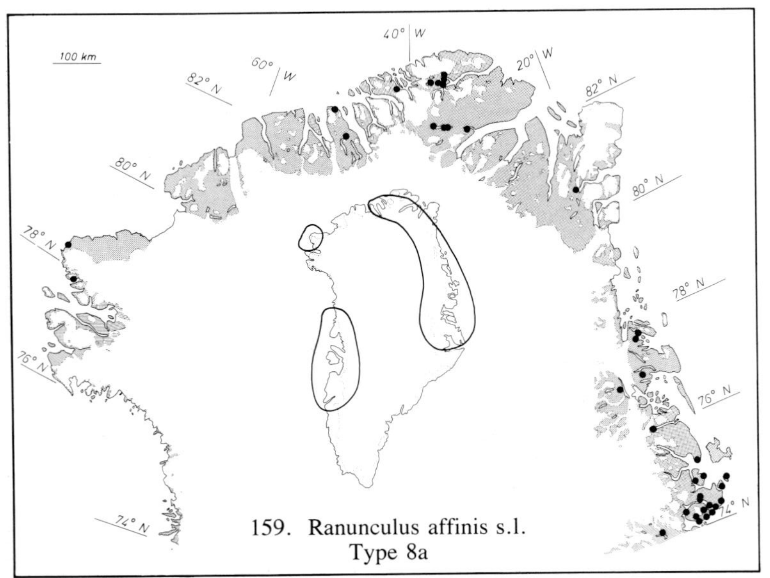

159. Ranunculus affinis s.l.
Type 8a

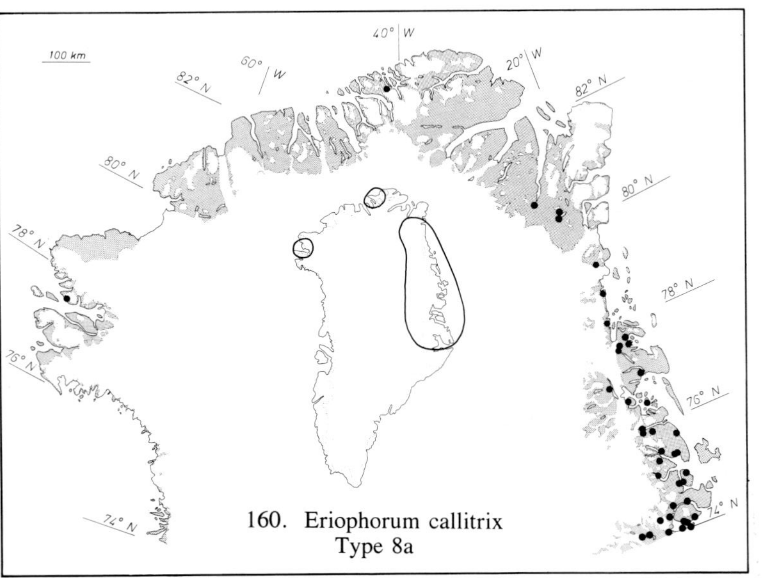

160. Eriophorum callitrix
Type 8a

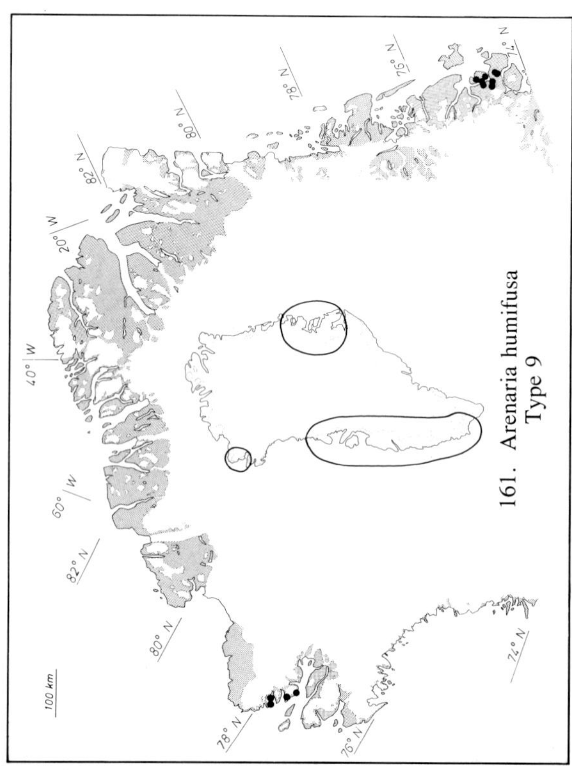

161. Arenaria humifusa
Type 9

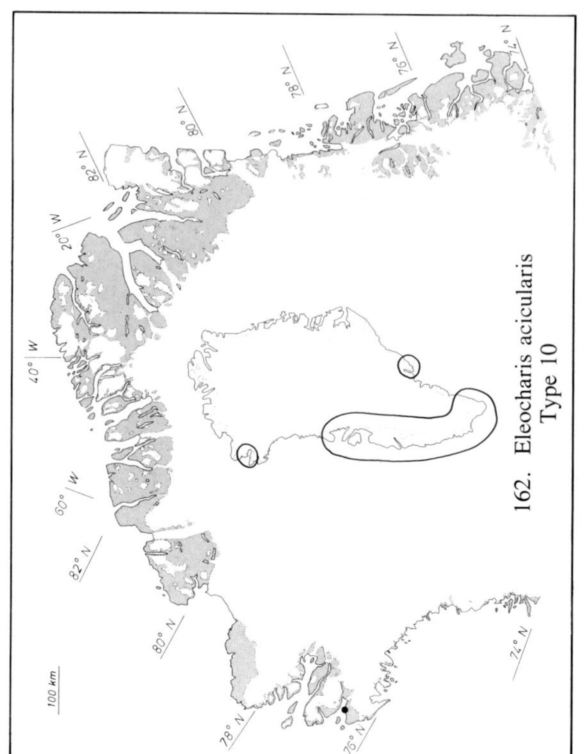

162. Eleocharis acicularis
Type 10

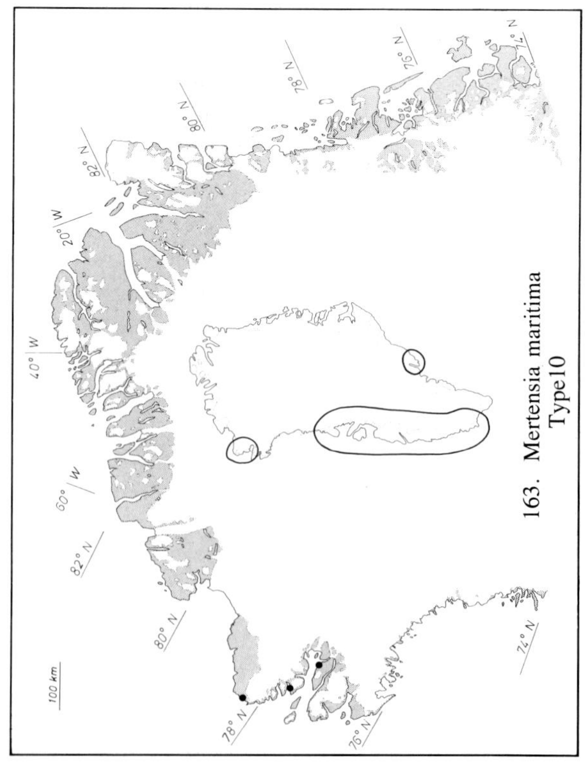

163. Mertensia maritima
Type10

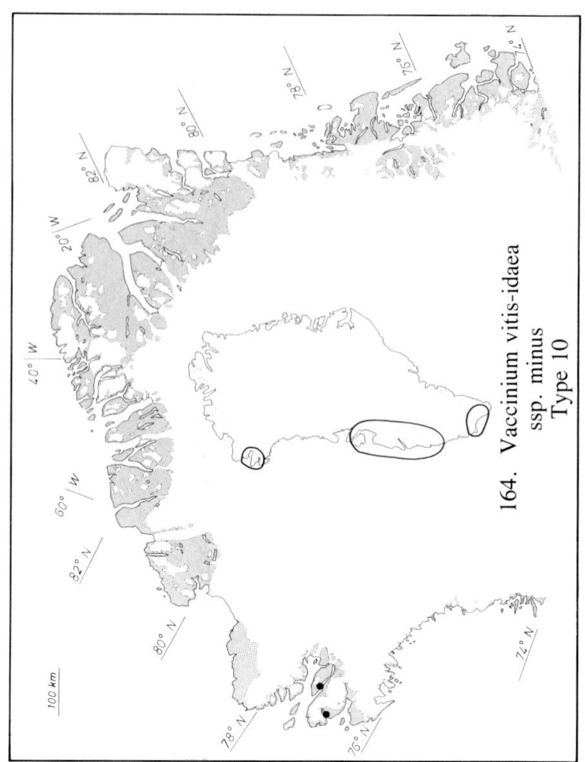

164. Vaccinium vitis-idaea
ssp. minus
Type 10

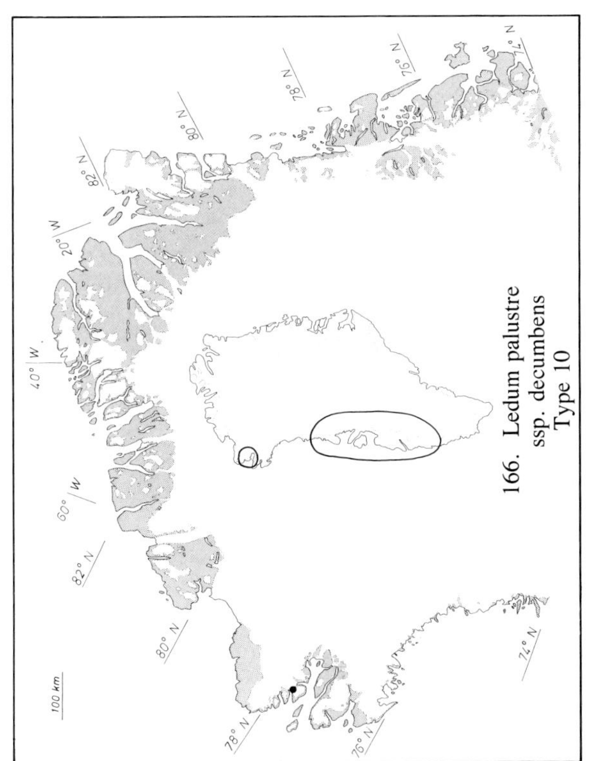

166. Ledum palustre ssp. decumbens
Type 10

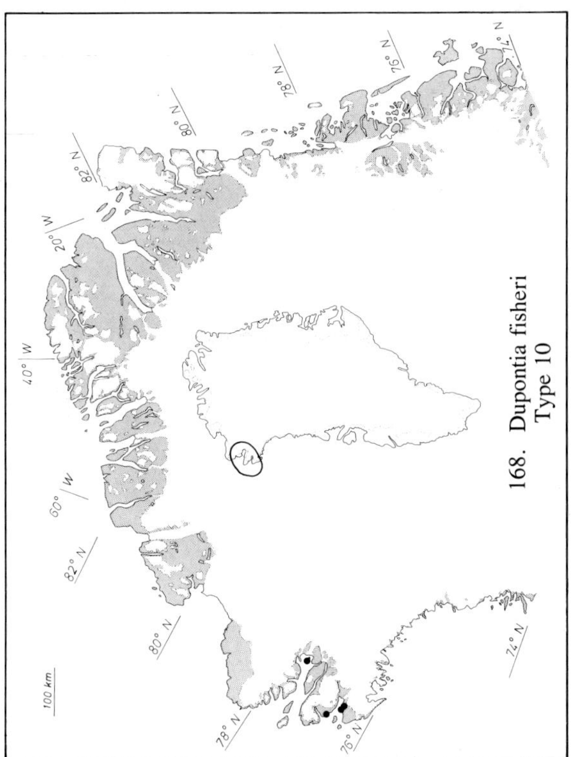

168. Dupontia fisheri
Type 10

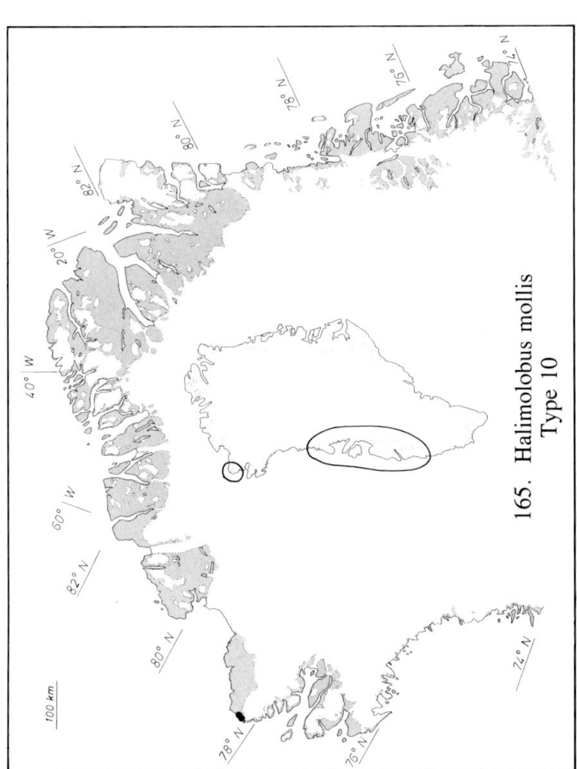

165. Halimolobus mollis
Type 10

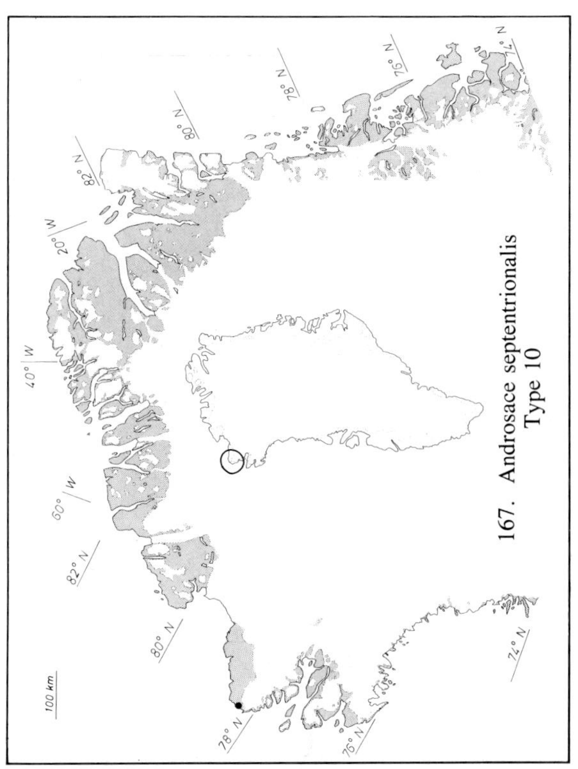

167. Androsace septentrionalis
Type 10

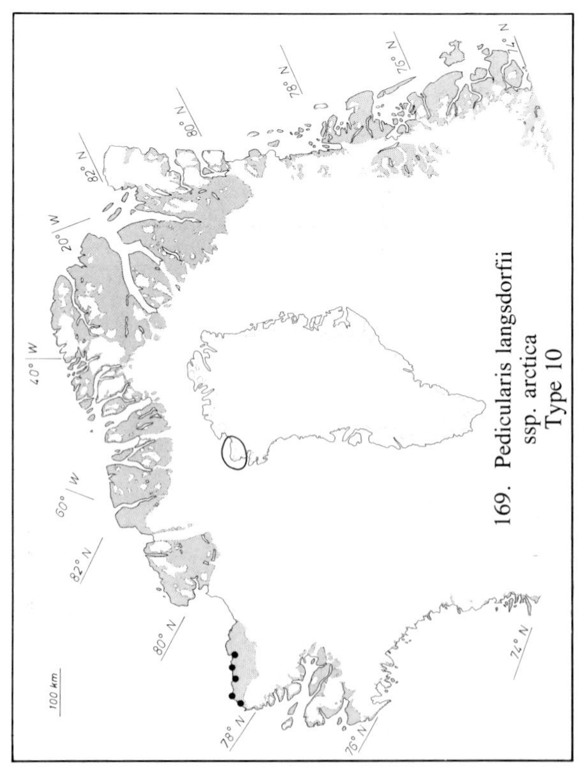

169. Pedicularis langsdorfii ssp. arctica
Type 10

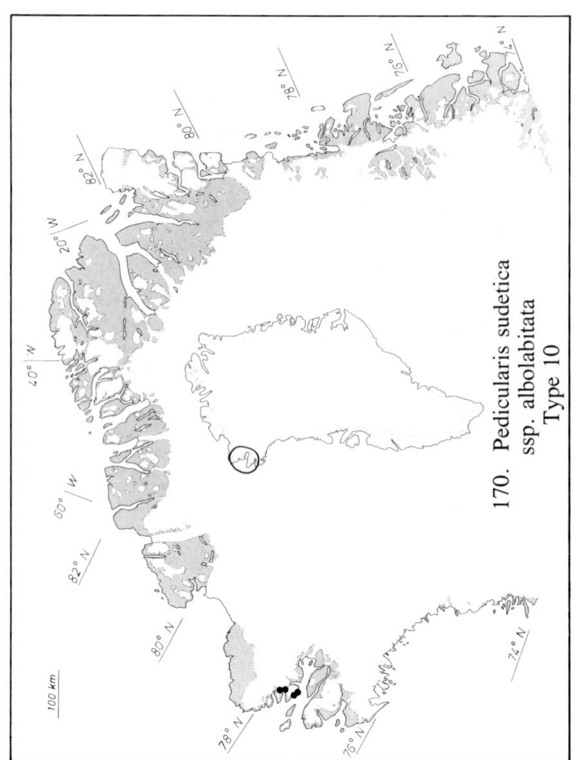

170. Pedicularis sudetica ssp. albolabitata
Type 10

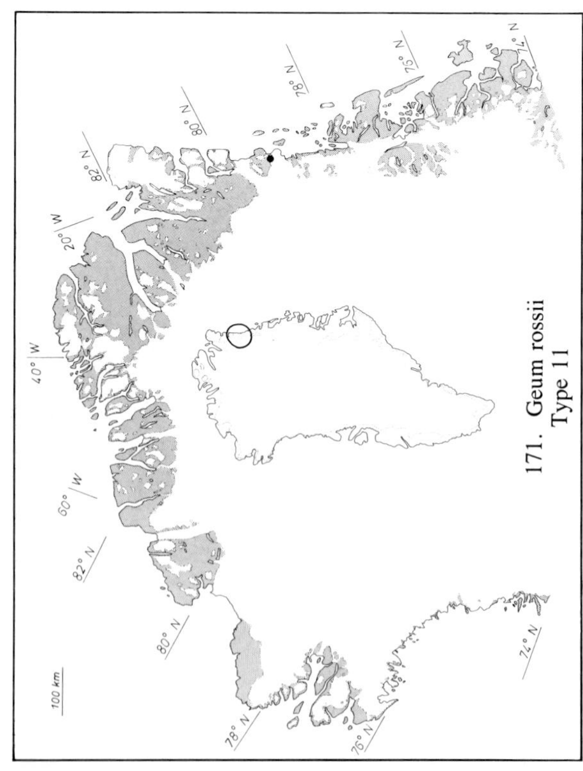

171. Geum rossii
Type 11

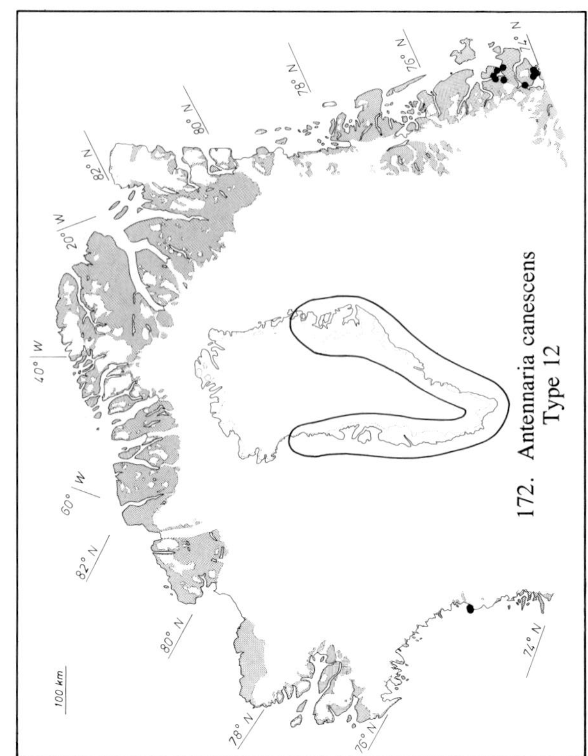

172. Antennaria canescens
Type 12

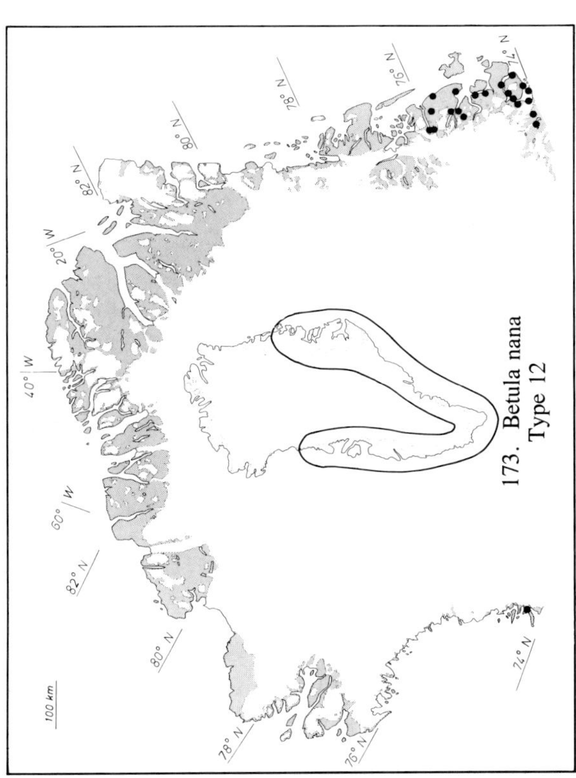

173. Betula nana
Type 12

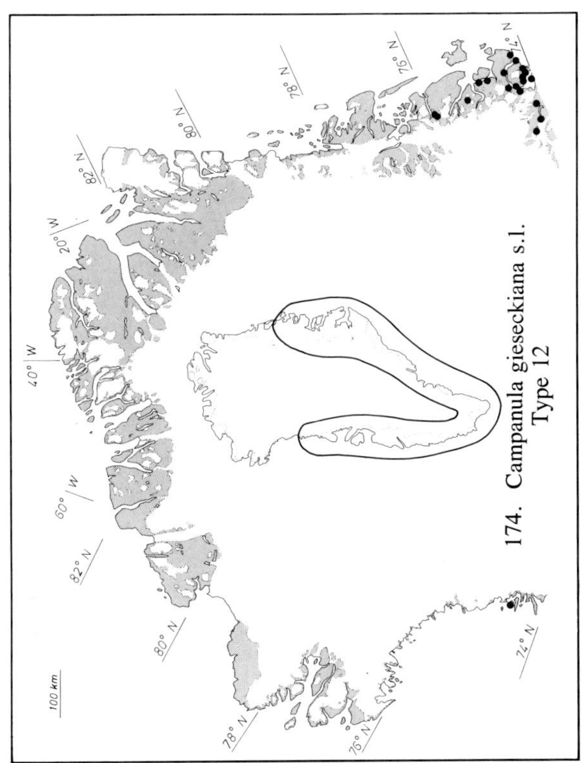

174. Campanula gieseckiana s.l.
Type 12

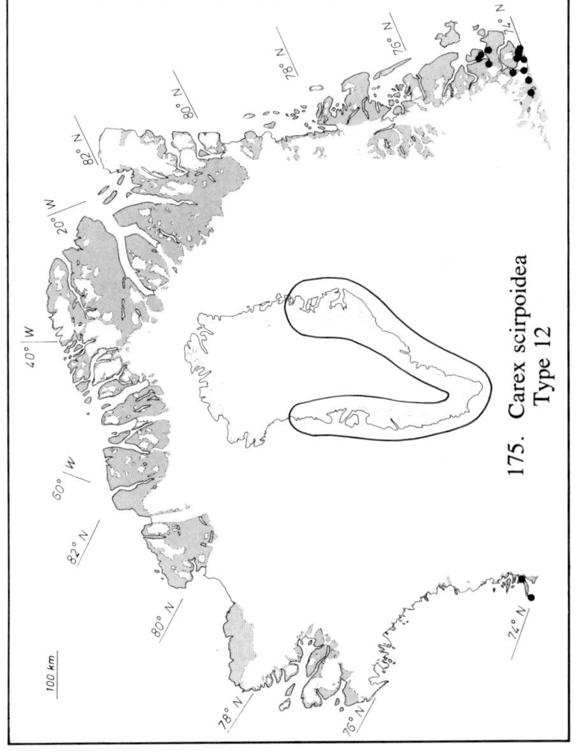

175. Carex scirpoidea
Type 12

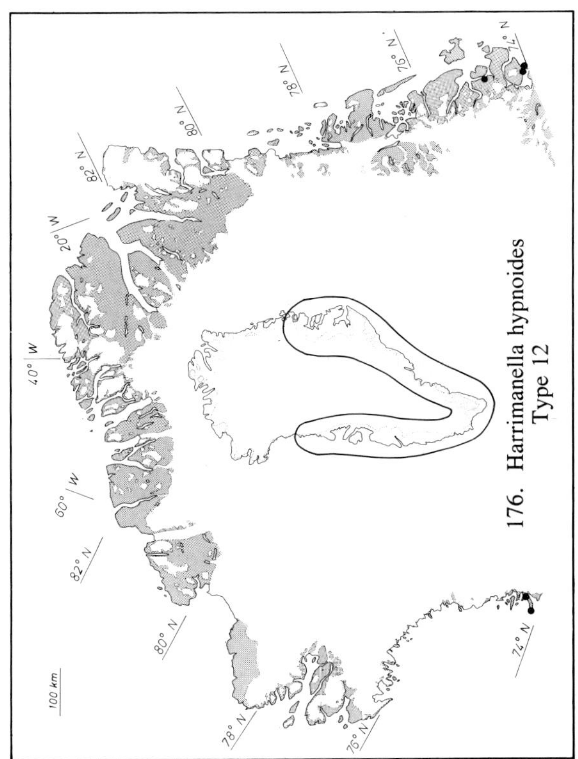

176. Harrimanella hypnoides
Type 12

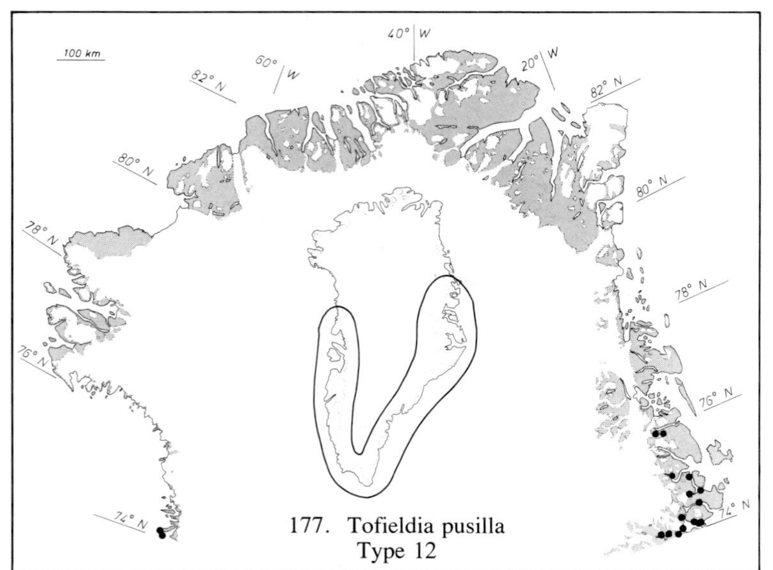

177. Tofieldia pusilla
Type 12

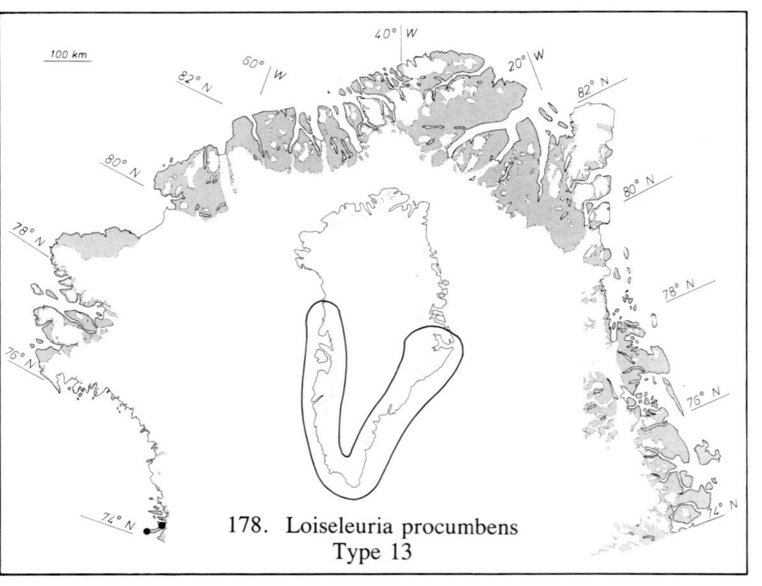

178. Loiseleuria procumbens
Type 13

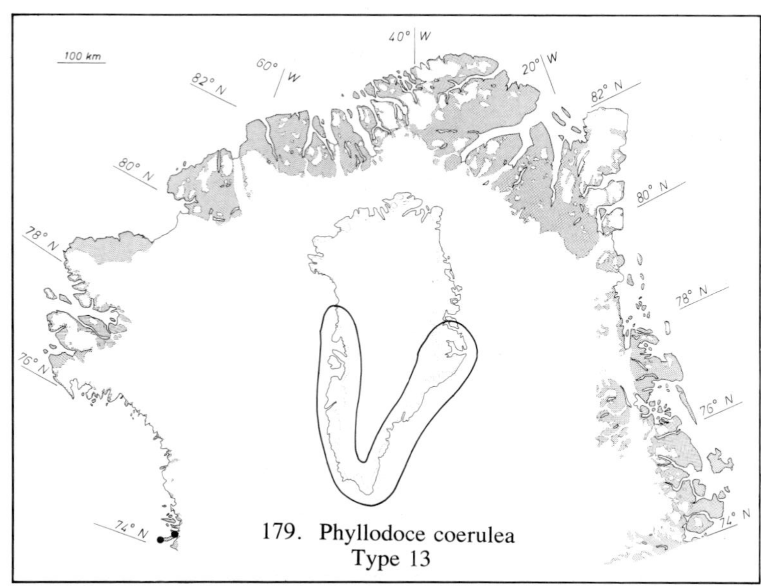

179. Phyllodoce coerulea
Type 13

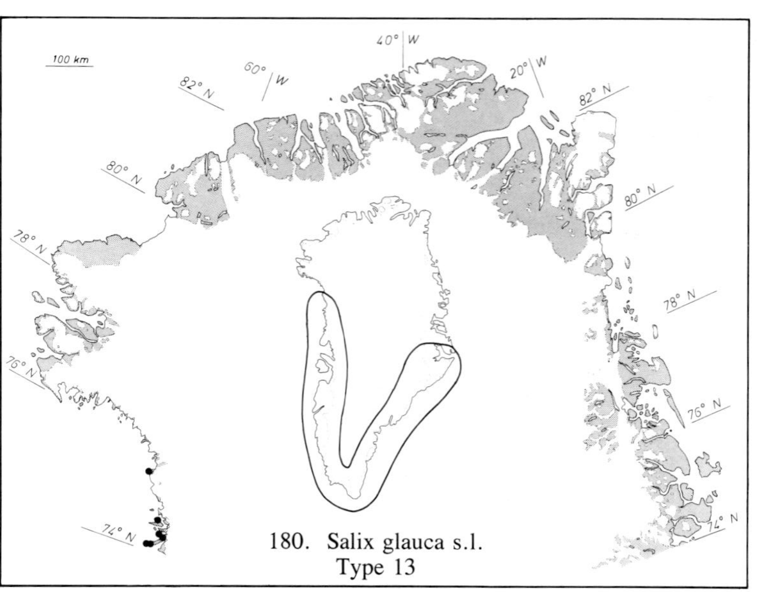

180. Salix glauca s.l.
Type 13

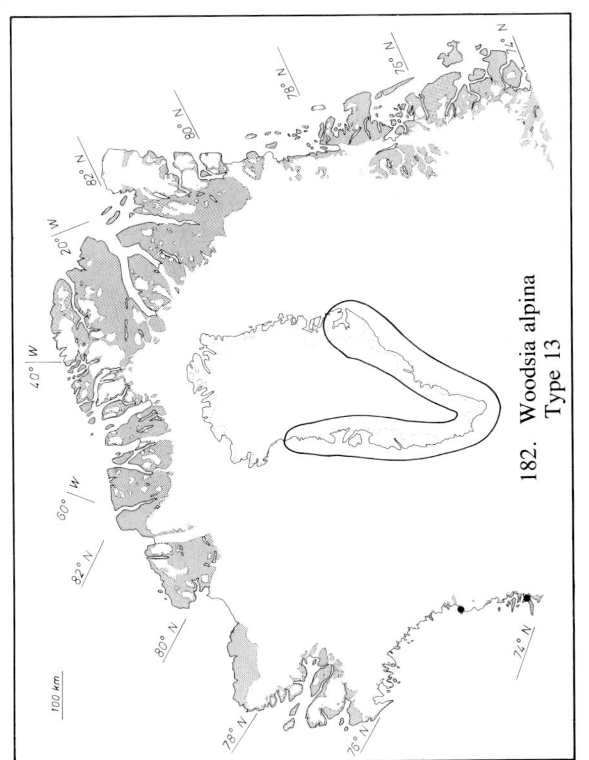

182. Woodsia alpina
Type 13

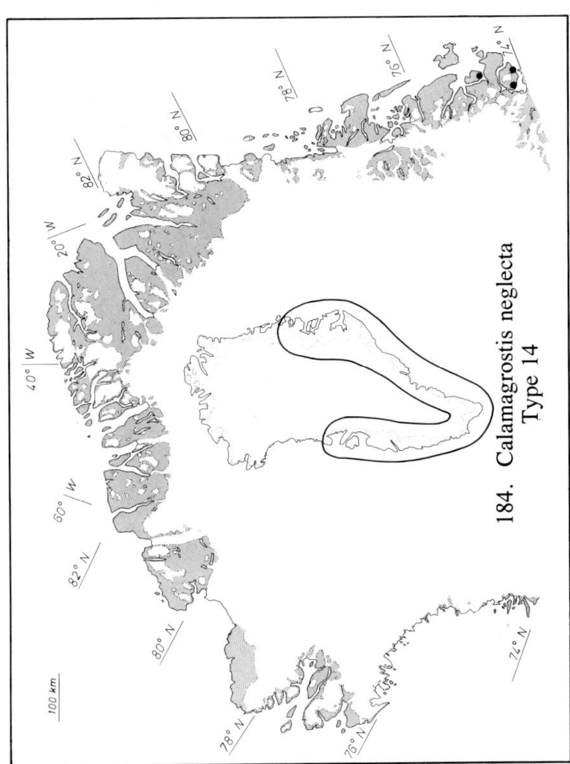

184. Calamagrostis neglecta
Type 14

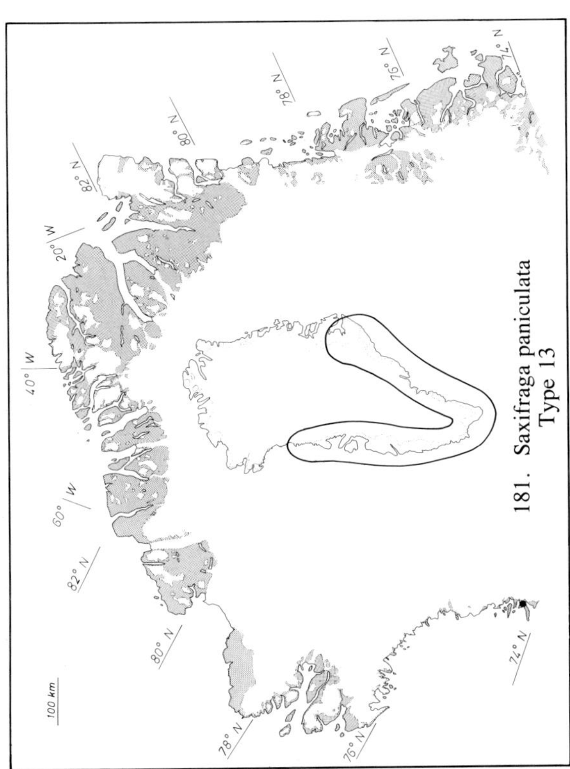

181. Saxifraga paniculata
Type 13

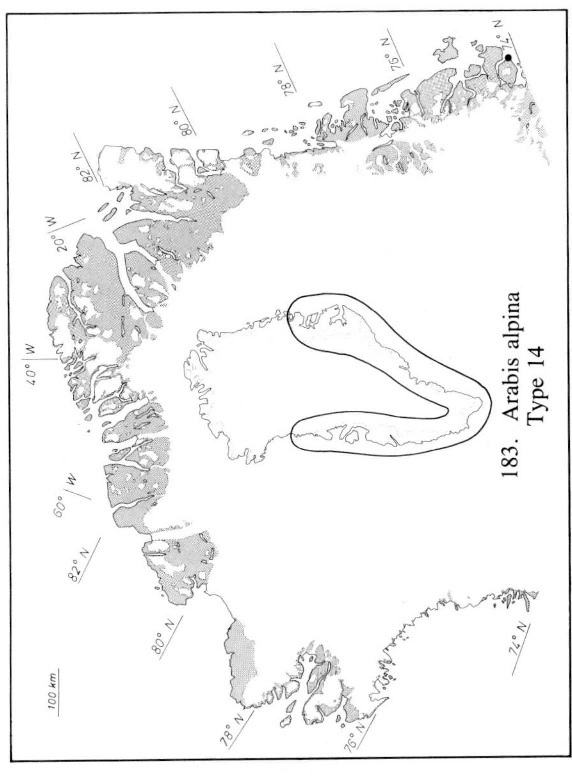

183. Arabis alpina
Type 14

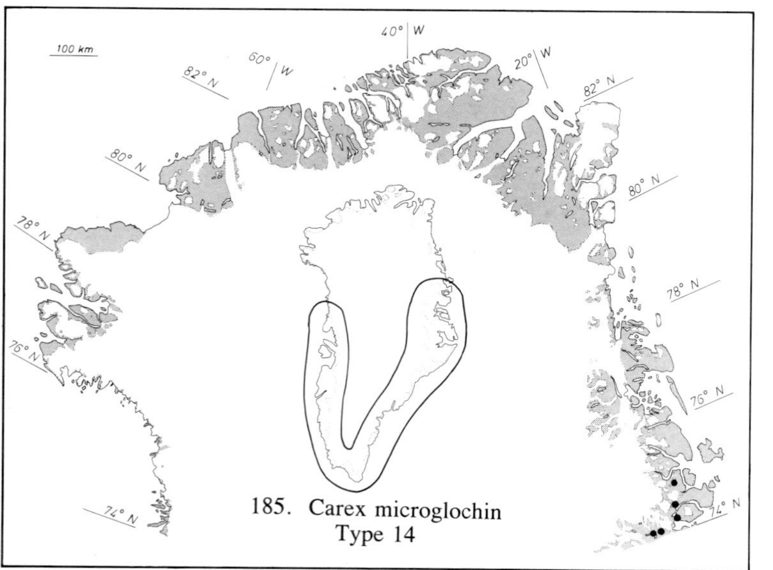

185. Carex microglochin
Type 14

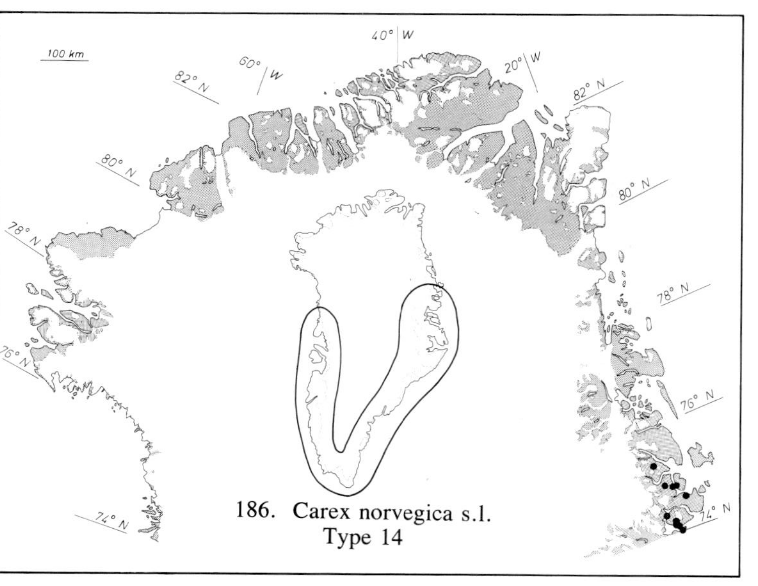

186. Carex norvegica s.l.
Type 14

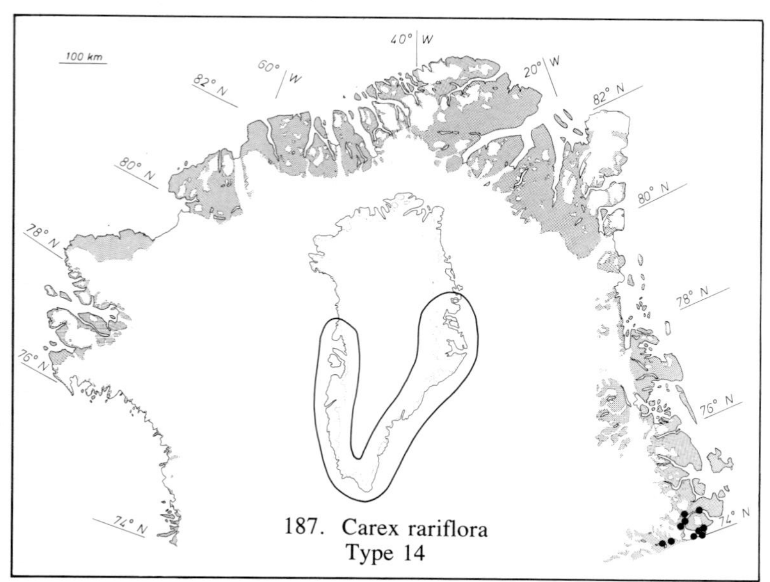

187. Carex rariflora
Type 14

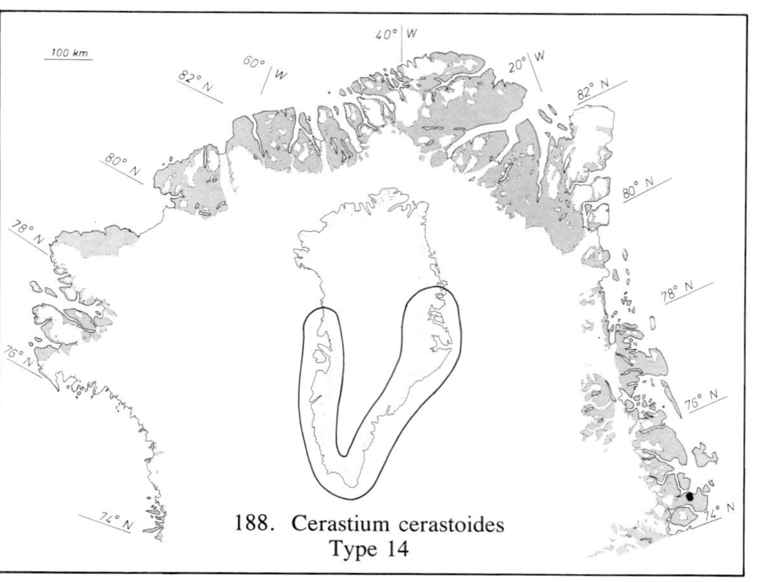

188. Cerastium cerastoides
Type 14

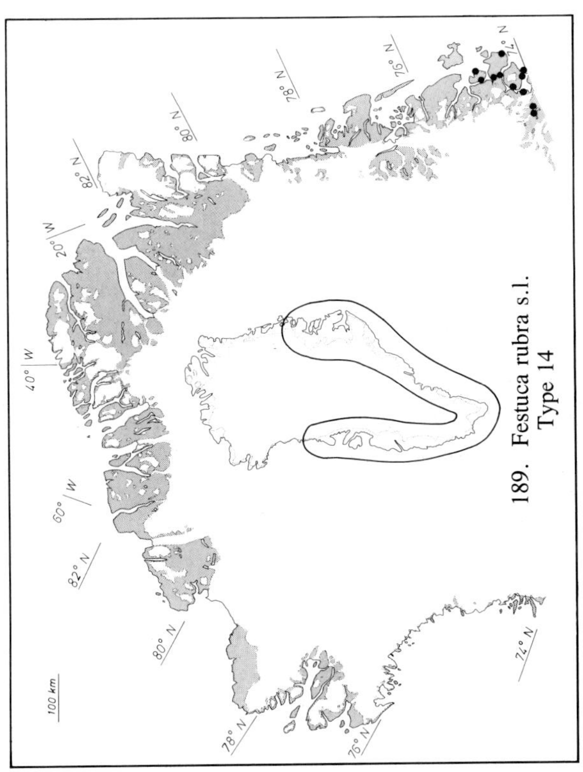

189. Festuca rubra s.l.
Type 14

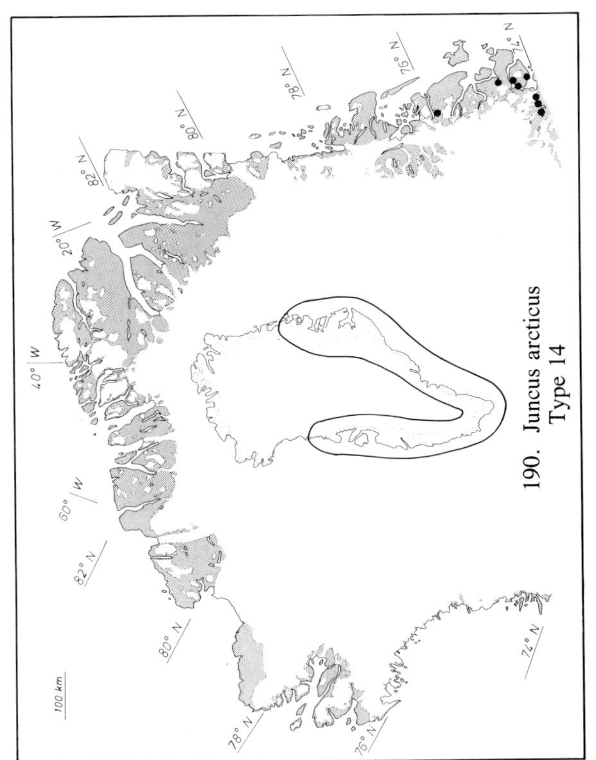

190. Juncus arcticus
Type 14

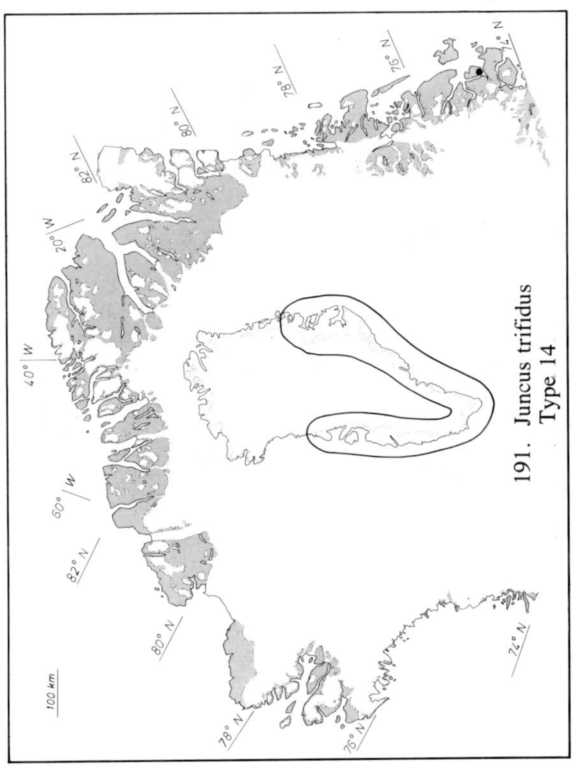

191. Juncus trifidus
Type 14

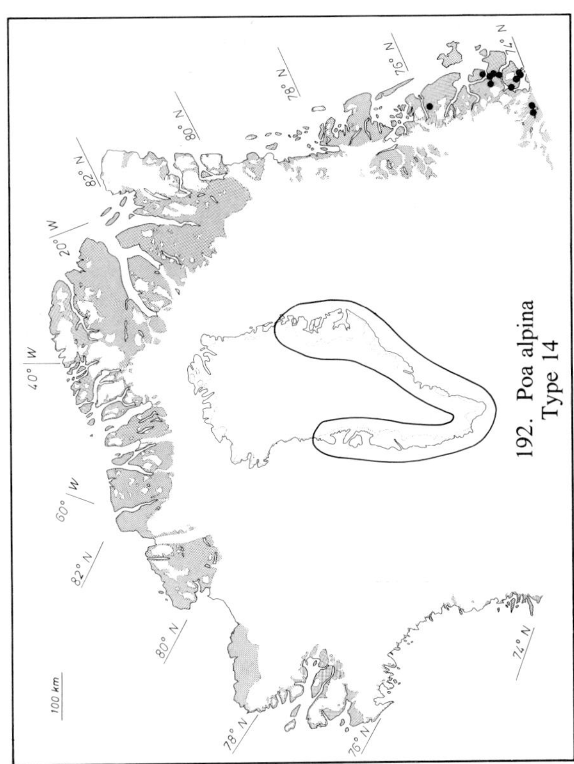

192. Poa alpina
Type 14

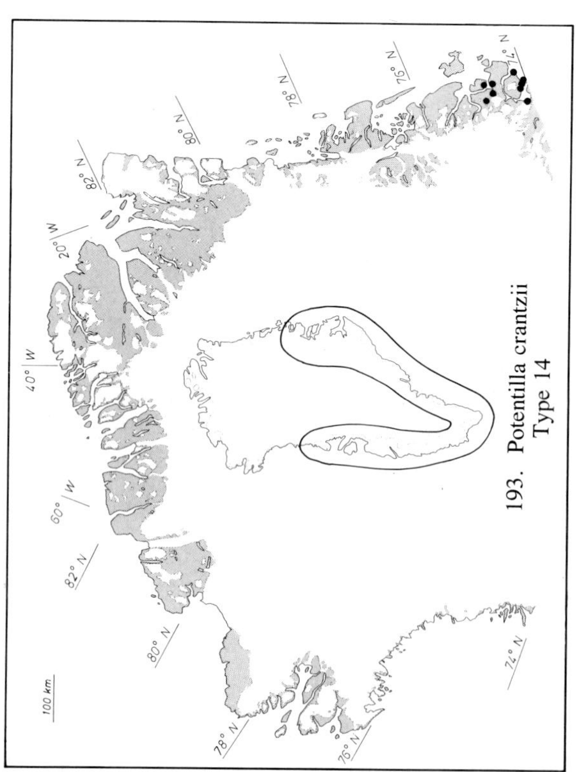

193. Potentilla crantzii
Type 14

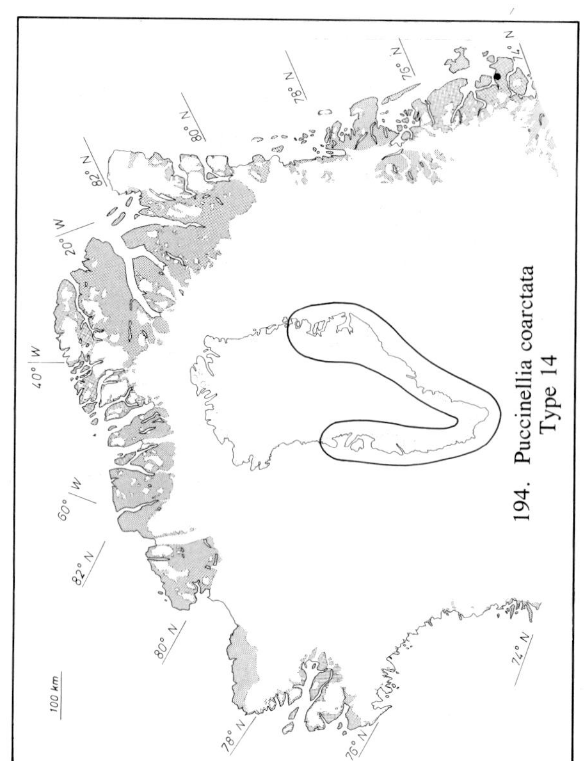

194. Puccinellia coarctata
Type 14

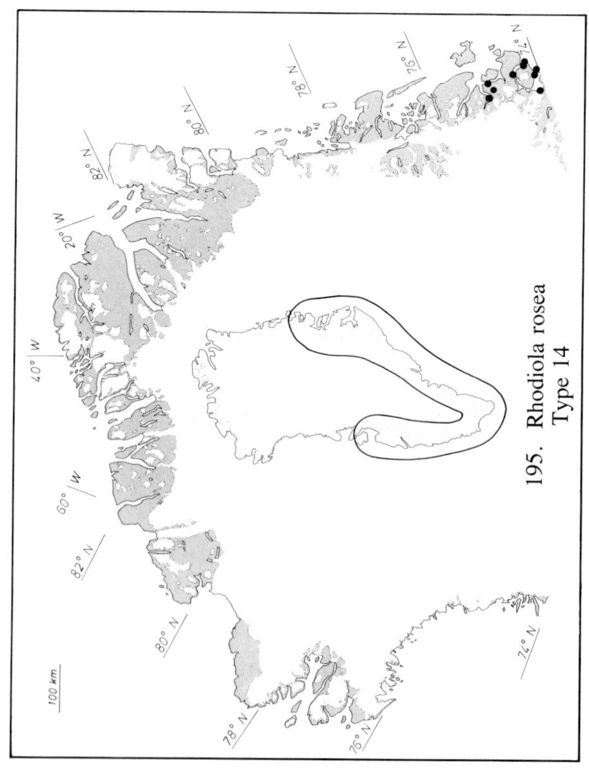

195. Rhodiola rosea
Type 14

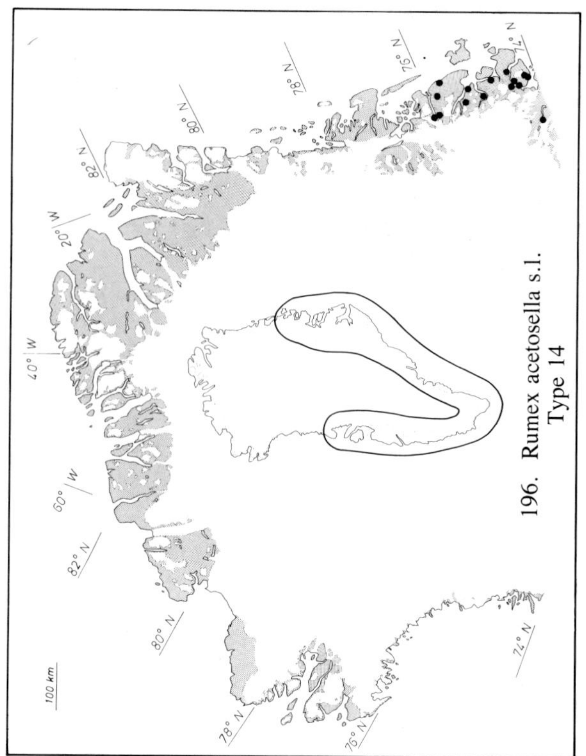

196. Rumex acetosella s.l.
Type 14

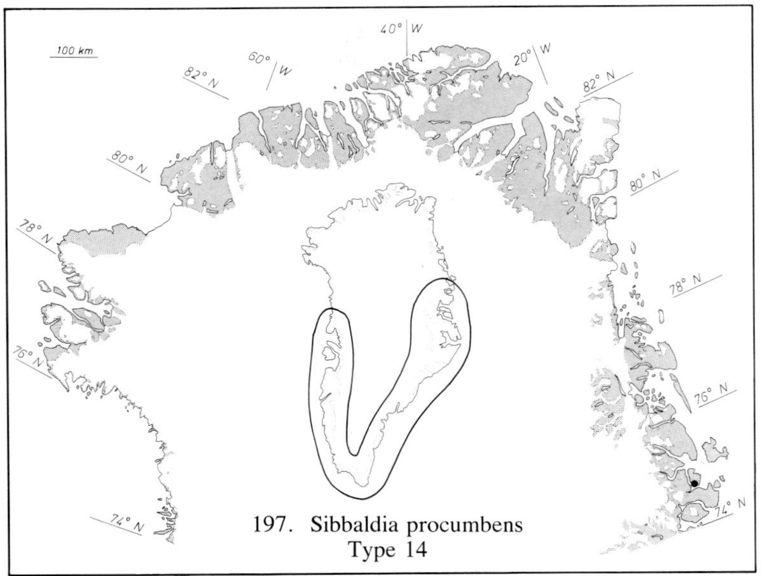

197. Sibbaldia procumbens
Type 14

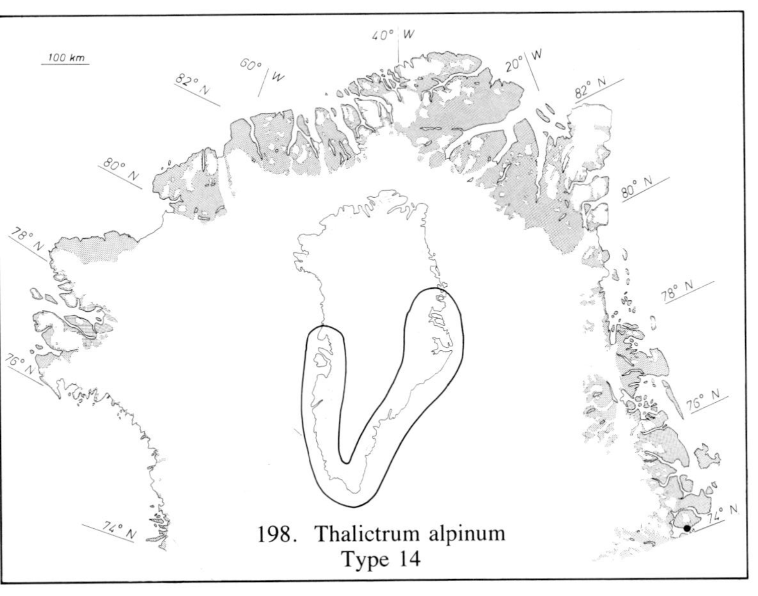

198. Thalictrum alpinum
Type 14

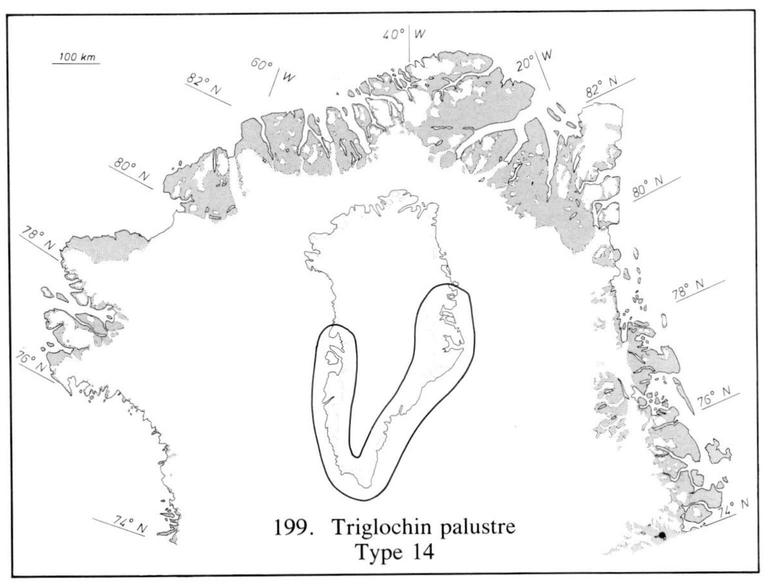

199. Triglochin palustre
Type 14

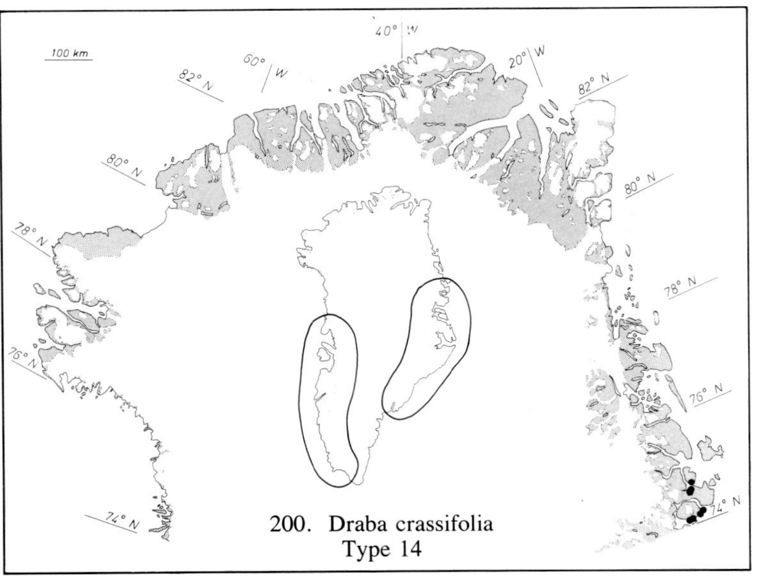

200. Draba crassifolia
Type 14

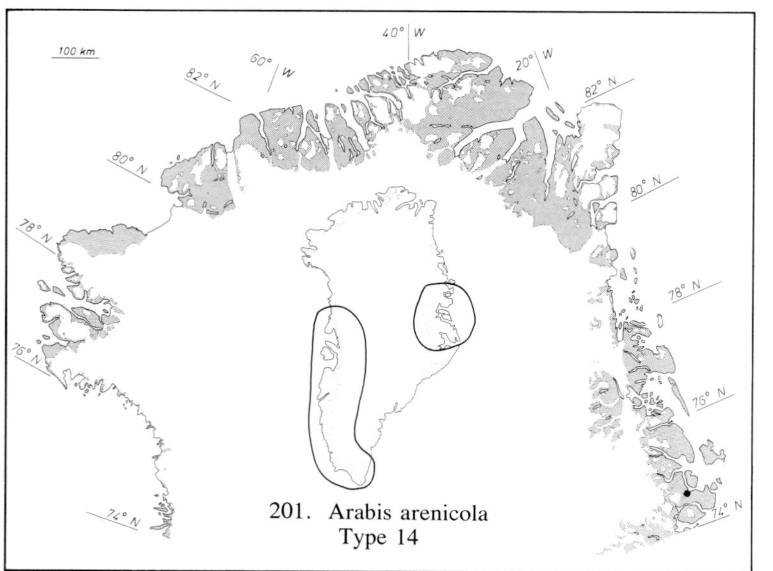
201. Arabis arenicola
Type 14

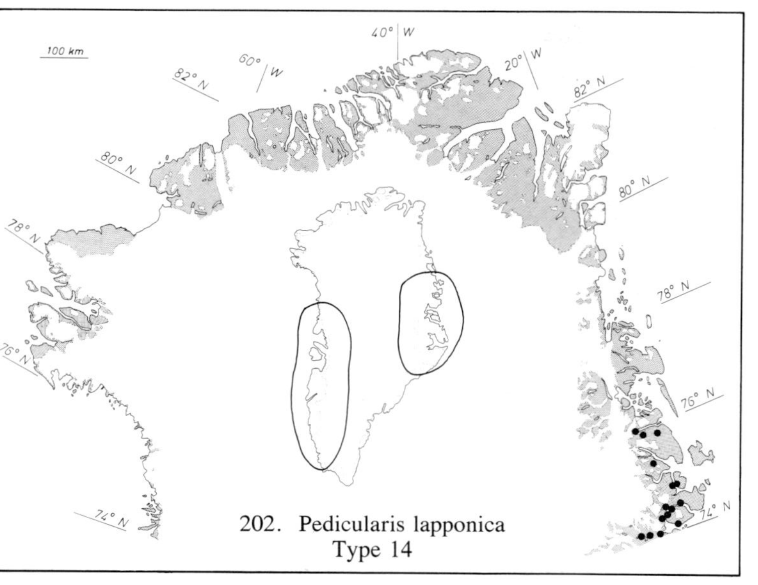
202. Pedicularis lapponica
Type 14

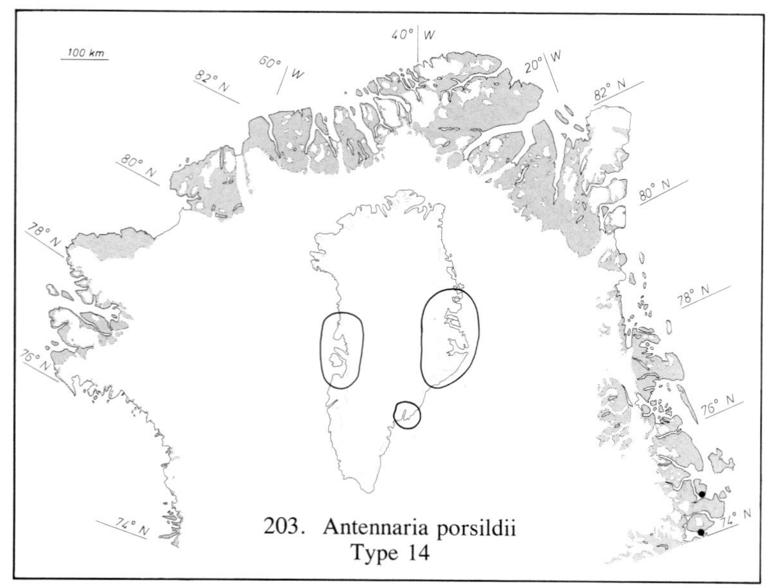
203. Antennaria porsildii
Type 14

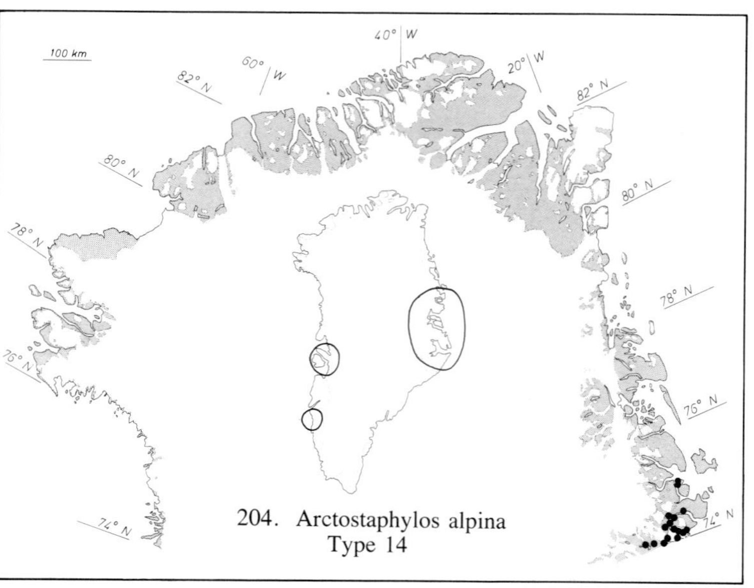
204. Arctostaphylos alpina
Type 14

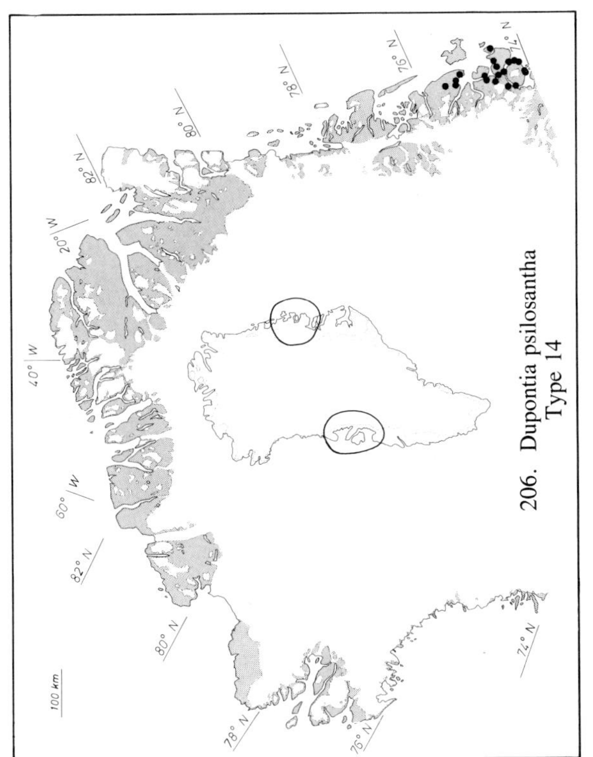

206. Dupontia psilosantha
Type 14

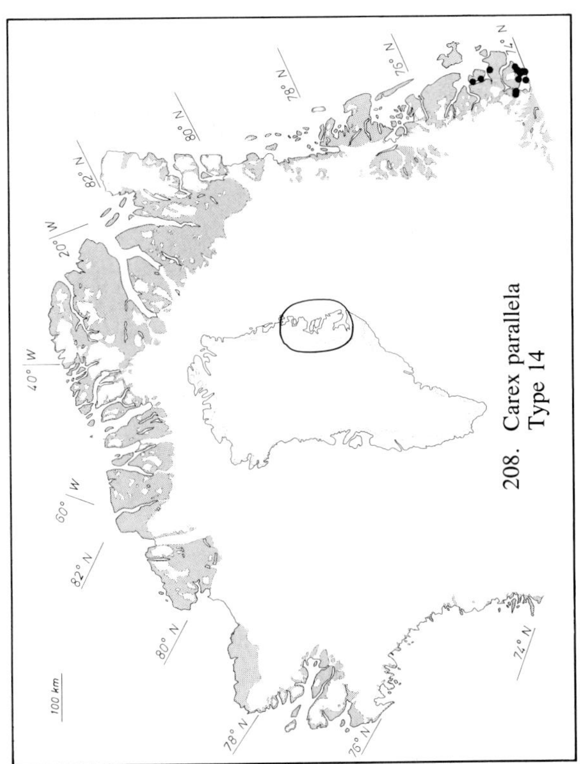

208. Carex parallela
Type 14

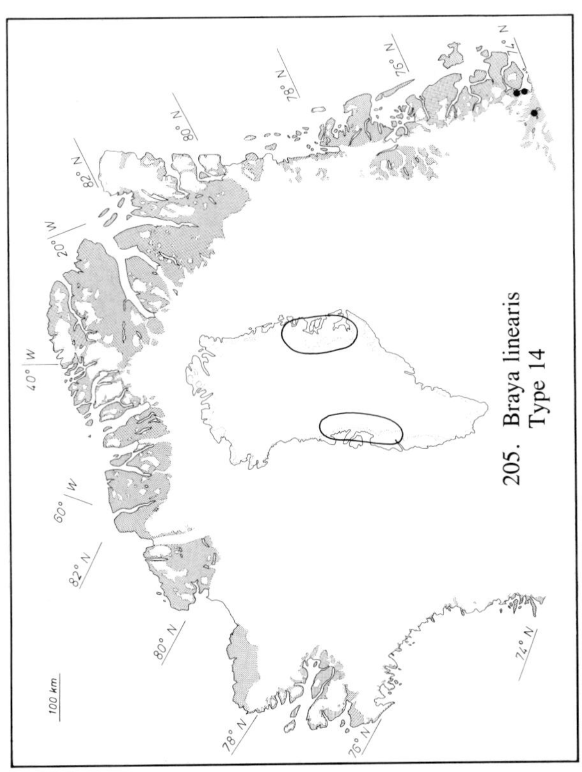

205. Braya linearis
Type 14

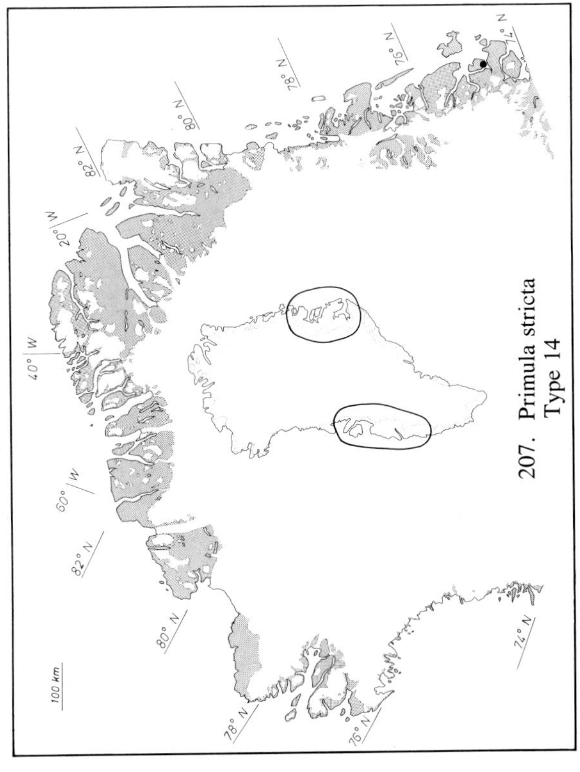

207. Primula stricta
Type 14

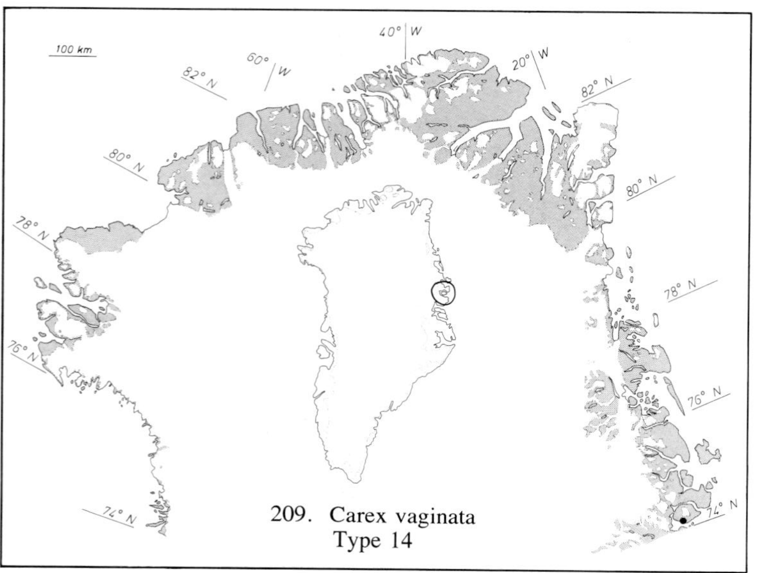

209. Carex vaginata
Type 14

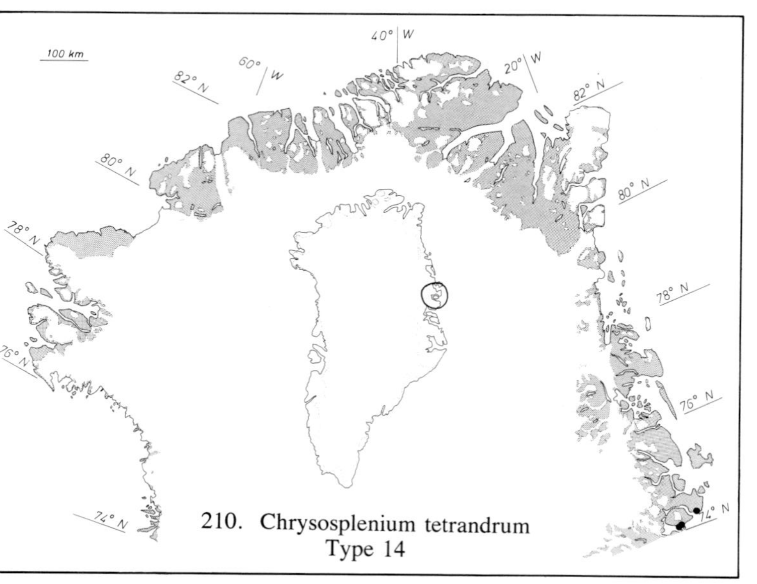

210. Chrysosplenium tetrandrum
Type 14

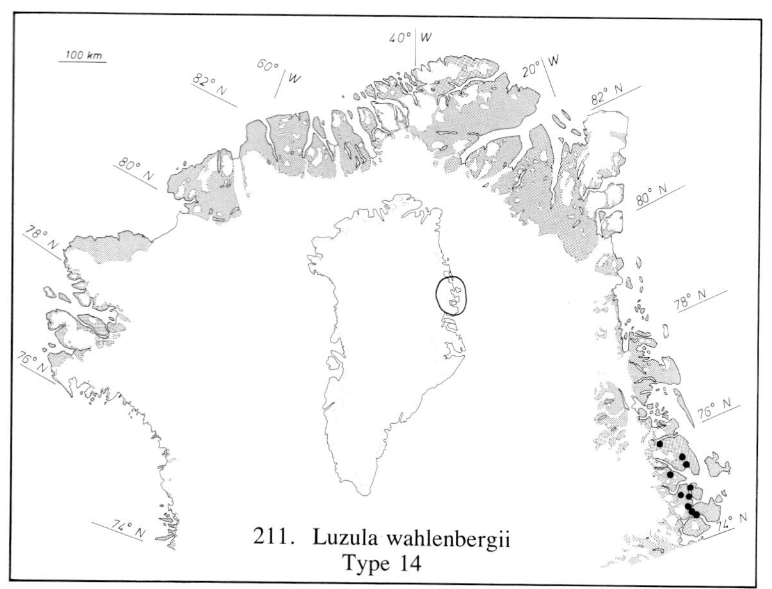

211. Luzula wahlenbergii
Type 14

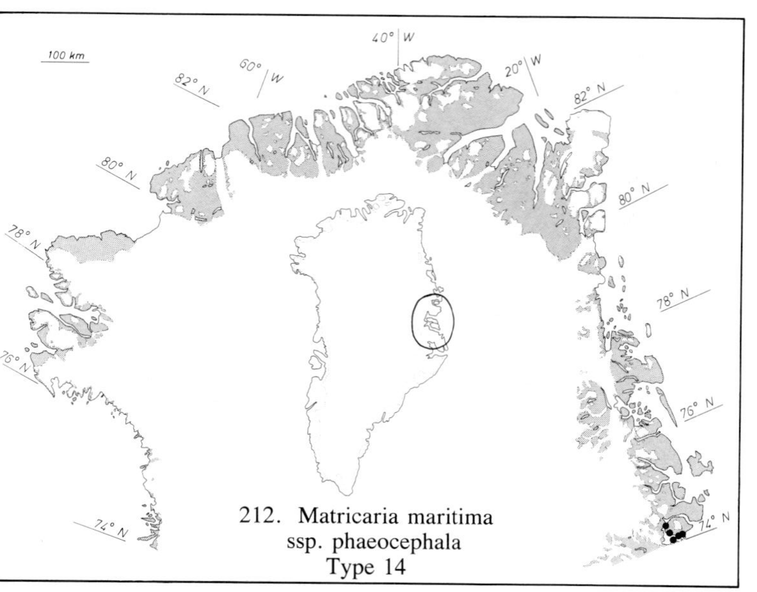

212. Matricaria maritima
ssp. phaeocephala
Type 14

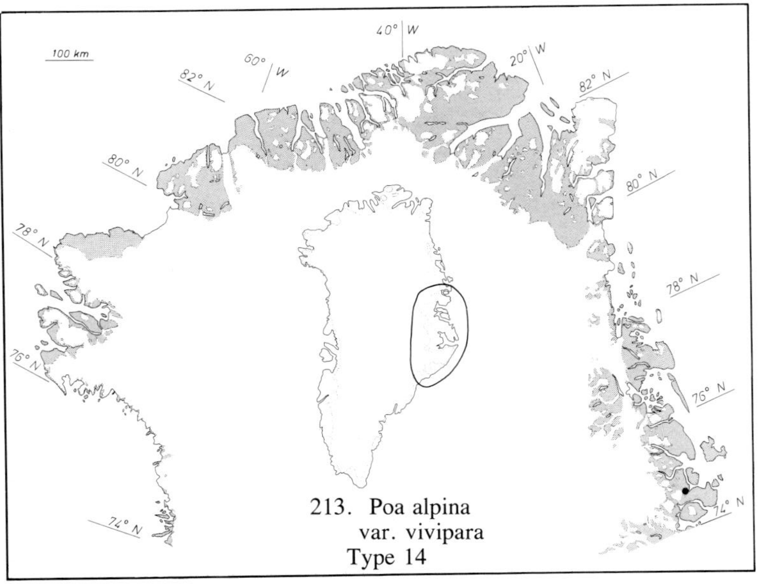

213. Poa alpina
var. vivipara
Type 14

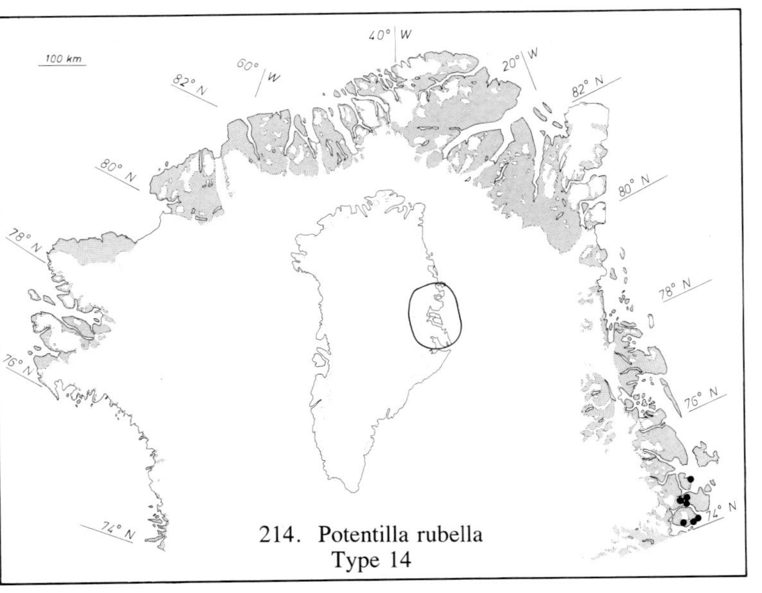

214. Potentilla rubella
Type 14

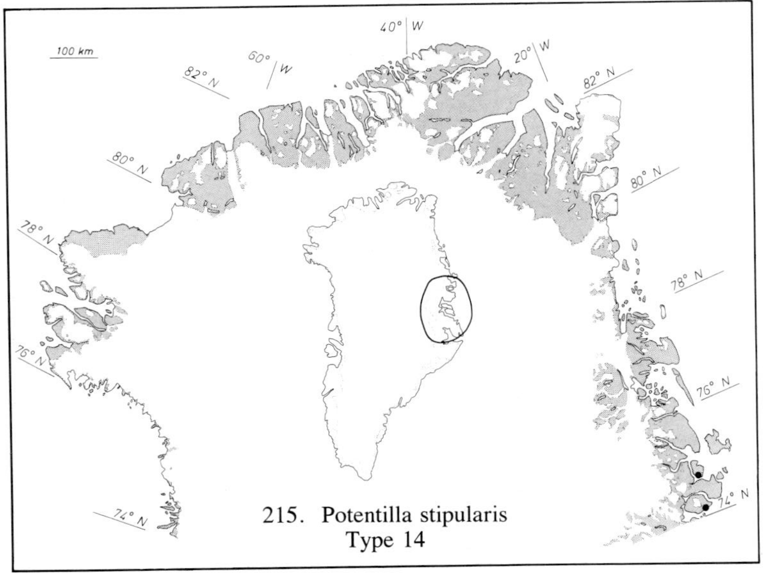

215. Potentilla stipularis
Type 14

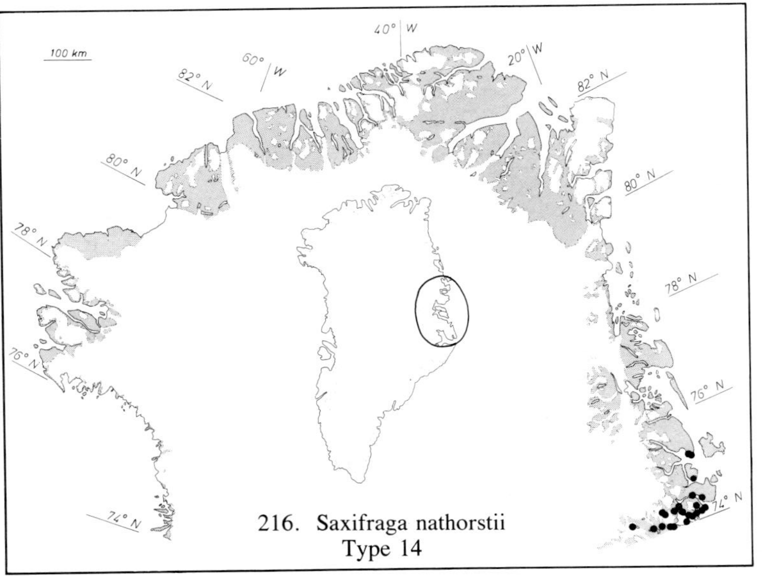

216. Saxifraga nathorstii
Type 14

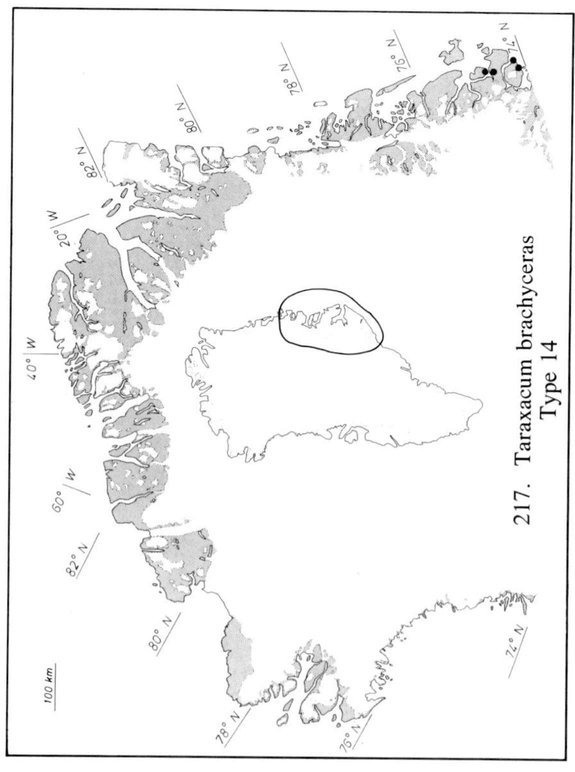

217. Taraxacum brachyceras
Type 14

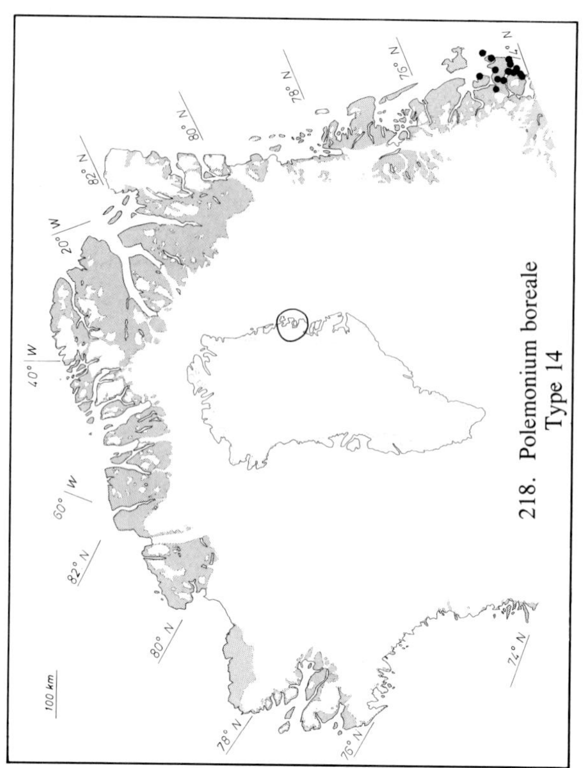

218. Polemonium boreale
Type 14

Instructions to authors

Two copies of the manuscript, each complete with illustrations, tables, captions, etc. should be sent to the Secretary, Kommissionen for videnskabelige Undersøgelser i Grønland. Manuscripts will be forwarded to referees for evaluation. Authors will be notified as quickly as possible about acceptance, rejection, or desired alterations. The final decision on these matters rests with the editor.

Manuscripts corresponding to less than 16 printed pages (of 6100 type units) including illustrations are not accepted, unless they are part of a special theme issue. Manuscripts that are long in relation to their content will not be accepted without abridgement.

Manuscript

Language. – Manuscripts should be in English (preferred language), French, or German. Authors who are not writing in their native language must have the language of their manuscript corrected before submission.

Place names. – All Greenland place names used in the text and in illustrations must be names authorised by The Greenlandic Language Committee. Authors are advised to submit sketch-maps with all required names to the Secretary for checking before the manuscript is submitted. Names of Greenland localities outside the area with which the paper is concerned should be accompanied by co-ordinates (longitude and latitude).

Title. – Titles should be as short as possible, with emphasis on words useful for indexing and information retrieval.

Abstract. – An abstract in English must accompany all papers. It should be short (no longer than 250 words), factual, and stress new information and conclusions.

Typescript. – Typescripts must be clean and free of handwritten corrections. Use double spacing throughout, and leave a 4 cm wide margin on the left-hand side. Avoid as far as possible dividing words at the right-hand end of a line. Consult a recent issue for general lay-out.

Page 1 should contain 1) title, 2) name(s) of author(s), 3) abstract, 4) key words (max. 10), 5) author's full postal address(es). Manuscripts should be accompanied by a table of contents, typed on separate sheet(s).

Underlining should only be used in generic and species names. The use of italics in other connections can be indicated by a wavy line in pencil under the appropriate words.

Use at most three grades of headings, but do not underline. The grade of heading can be indicated in soft pencil in the left hand margin of one copy of the typescript. Avoid long headings.

Floppy disc. – It may be helpful in the printing procedure if, in addition to the hard copies, the manuscript is also submitted on a DOS-formatted floppy disc. However, editing will be made on the hard copy, and the text file on the disc must be identical to the final version of the manuscript.

References. – References to figures and tables in the text should have the form: Fig. 1, Figs 2–4, Table 3. Bibliographic references in the text are given thus: Shergold (1975: 16) ... (Jago & Daily 1974b).

In the list of references the following style is used:

Boucot, A. J. 1975. Evolution and extinction rate controls. – Elsevier, Amsterdam: 427 pp.

Sweet, W. C. & Bergström, S. M. 1976. Conodont biostratigraphy of the Middle and Upper Ordovician of the United States midcontinent. – In: Bassett, M. G. (ed.). The Ordovician System: Proceedings of a Palaeontological Association symposium, Birmingham, September 1974: 121–151. University of Wales Press.

Tarling, D. H. 1967. The palaeomagnetic properties of some Tertiary lavas from East Greenland. – Earth and Planetary Science Letters 3: 81–88.

Meddelelser om Grønland, Bioscience (Geoscience, Man & Society) should be abbreviated thus: *Meddr Grønland, Biosci. (Geosci., Man & Soc.).*

Illustrations

General. – Submit two copies of all diagrams, maps, photographs, etc., all marked with number and author's name. Normally all illustrations will be placed in the text.

All figures (including line drawings) must be submitted as glossy photographic prints suitable for direct reproduction, and preferably have the dimensions of the final figure. Do not submit original artwork. Where appropriate the scale should be indicated on the illustration or in the caption.

The size of the smallest letters in illustrations should not be less than 1.3 mm. Intricate tables are often more easily reproduced as text figures than by type-setting; when lettering such tables use "Letraset" or a typewriter with carbon ribbon.

Colour plates may be included at the author's expense, but the editor must be consulted before such illustrations are submitted.

Size. – The width of figures must be that of a column (76.5 mm), 1½ columns (117 mm), or a page (157 mm). The maximum height of a figure (including caption) is 217 mm. Horizontal figures are preferred. If at all possible, fold-out figures and tables should be avoided.

Caption. – Captions to figures must be typed on a separate sheet and submitted, like everything else, in duplicate.

Proofs

Authors receive two page proofs. Prompt return to the editor is requested. Only typographic errors should be corrected in proof; the cost of making alterations to the text and figures at this stage will be charged to the author(s).

Twenty-five copies of the publication are supplied free, fifty if there are two or more authors. Additional copies can be supplied at 55% of the retail price. Manuscripts (including illustrations) are not returned to the author after printing unless specifically requested.

Copyright

Copyright in all papers published by Kommissionen for videnskabelige Undersøgelser i Grønland is vested in the Commission. Those who ask for permission to reproduce material from the Commission's publications are, however, informed that the author's permission must also be obtained if he is still alive.

8763512173

9 788763 512176

Meddelelser om Grønland

Bioscience
Geoscience
Man & Society

Published by
The Commission
for Scientific
Research
in Greenland

Printed in Denmark by AiO Print as, Odense